Numerical Methods for Diffusion Phenomena in Building Physics

Nathan Mendes · Marx Chhay ·
Julien Berger · Denys Dutykh

Numerical Methods for Diffusion Phenomena in Building Physics

A Practical Introduction

Nathan Mendes
Thermal Systems Laboratory (LST)
Pontifical Catholic University of Paraná
Curitiba, Brazil

Marx Chhay
Université Savoie Mont Blanc
Université Grenoble Alpes
Chambéry, France

Julien Berger
CNRS, LOCIE
Université Grenoble Alpes
Chambéry, France

Denys Dutykh
Laboratoire de Mathématiques
Université Savoie Mont Blanc
Le Bourget-du-Lac, France

ISBN 978-3-030-31576-4   ISBN 978-3-030-31574-0   (eBook)
https://doi.org/10.1007/978-3-030-31574-0

Original edition published by © Editora Universitária Champagnat (PUCPRESS), Curitiba, Brazil in 2016. ISBN 978-85-68324-45-5
© Springer Nature Switzerland AG 2019

This Springer imprint is published by the registered company Springer Nature Switzerland AG
The registered company address is: Gewerbestrasse 11, 6330 Cham, Switzerland

For my daughters Clara and Laura,
For my parents.
—NM

Para minha Jolane,
À mon père pour tout.
—MC

À Maurice et Thérèze,
À ma famille.
—JB

For my wife Katya,
For my parents.
—DD

Foreword

It is with great pleasure that I have read this book! It represents a very ambitious step, suggesting future improvements in numerical techniques used in the field of building physics today.

Building physics deals with the physics (heat, air, and moisture) of the building envelope and indoor environment. The couplings to the external environment and human activities drive the processes. Here, both the local climatic conditions and the surrounding human settlements are of importance.

To solve the problems at hand, numerical methods that can solve nonlinear and coupled problems containing heat and mass transfer are essential. In order to be used in practice, they also need to be reasonably efficient in terms of time and accuracy.

In the past decades, numerical models have developed considerably in the field of building physics, and are in line with the rapid development of computer capacity, and are in general user-friendly. Today, these models can analyze much more complex problems, incorporating more detailed material model descriptions as well as the interaction between different building elements and environmental conditions. However, there are still more challenges to be met.

This book begins with an overview of the past 30 years of research in this area. It emphasizes the important steps by individual researchers as well as collaborative international research in the field. The authors give the basis of today's finite-difference and finite-element methods used. They continue with reduced order, boundary integral, and spectral methods. These have previously only been used for particular problems and not incorporated in the mainstream models. The different models are described using very concise and compact mathematical formulations, and are explained through examples and exercises.

This book will be excellent as reference literature and a textbook. It will work very well in the fostering of new Ph.D. students for state-of-the-art numerical methods in building physics as well as inspiring them to make new developments in the field.

May 2018
Prof. Carl-Eric Hagentoft
Chalmers University of Technology
Göteborg, Sweden

Preface

Research in diffusion phenomena was started by brilliant scientists in the nineteenth century, and by the twentieth century, this phenomenon became widespread in many areas such as hydrology, agriculture, atmospheric sciences, geophysics, environmental engineering, food science, energy systems, and building physics, where the transient evolution of energy and mass fields plays a major role and is closely interrelated with particular physical characteristics of porous microstructures and boundary conditions subject to the prevailing environmental conditions.

Different models express diffusion down a concentration gradient according to the Fick's law, based on which complexity or simplicity can be improved depending on the nature of the diffused quantities and their interrelation. In the field of building physics, highly coupled important diffusion phenomena are present in terms of both heat and mass transfer processes through the buildings' porous envelope, roofing systems, ground, and furniture, which may be of a paramount importance to the building performance. However, in building energy tools, during the conduction of loads through porous walls, usually the storage and transport of moisture in the porous structure of the walls are neglected; these are normally submitted to both thermal and moisture gradients, so that an accurate heat transfer determination would require a simultaneous calculation of both sensible and latent effects in three dimensions by considering their highly nonlinear coupled terms. Besides its effect on the building energy performance, moisture has other implications, especially in humid climates. It is well known that moisture can cause damage to the building structure and can promote the growth of mold and mildew, thus affecting the health of building occupants.

In the past three decades, many researchers have developed models for moisture transport in porous materials. However, most models are still restricted to 1D transfer or are not combined into whole-building models. To solve diffusion problems by considering variable properties, dependent local sources, and moisture effects, some building simulation programs such as EnergyPlus, DoMUS, BSim, WuFI, and ESP-r adopt numerical methods to more precisely compute the energy and mass balances for each zone and for the whole building. Nevertheless, their approach is somehow restricted to some hypotheses, and thus further research and development are required in this field to provide more reliable predictions.

Therefore, this book intends to stimulate research in this area, by first providing an overview of some models and numerical techniques of the diffusion phenomena of heat, air, and moisture in porous building elements (Chap. 2). Next, a practical introduction on the finite-difference and finite-element methods traditionally used in building energy and hygrothermal simulation tools is provided (Chaps. 3 and 4). The second part of this book presents non-traditional methods, which are considered advanced in the sense that they have not been explored for solving diffusion phenomena in building physics. The advantage of these methods includes the improvement of the solution of heat and mass diffusion phenomena, especially in whole-building large domains relevant to building energy performance analysis. Therefore, the implementation of new numerical methods in the area, such as reduced order models, boundary integral approaches, and spectral methods, might be considered in the next generation of building-energy-simulation tools to more precisely and effectively predict energy consumption and demand, thermal comfort, mold growth risk, and material deterioration, considering for instance, the three-dimensional nature of the diffusive phenomena under asymmetric and heterogeneous boundary conditions. The third part of this book (Chap. 9) aims at providing some exercises to introduce students on the solution of diffusion phenomena in building physics.

The book chapters represent the content of lectures given at a Ph.D. International school entitled "Numerical Methods for Diffusion Phenomena in Building Physics: Theory and Practice", which took place at the Pontifical Catholic University of Parana (PUCPR, Curitiba–Brazil) in April 2016. The book is organized by the authors, and the chapters are written according to their specialty. Chapter 2 is produced by Prof. Nathan Mendes; Chaps. 3, 4, and 7 by Dr. Marx Chhay; Chap. 6 by Dr. Julien Berger; and the brief history of diffusion Chap. 1 and Chaps. 5, 8 are produced by Dr. Denys Dutykh.

Curitiba, Brazil   Nathan Mendes
Aix-les-Bains, France   Marx Chhay
Chambéry, France   Julien Berger
Chambéry, France   Denys Dutykh
May 2018

Acknowledgements

The authors acknowledge the Coordination for the Improvement of Higher Education Personnel (CAPES) within the Brazilian Ministry of Education and the French Committee for the Evaluation of Academic and Scientific Cooperation with Brazil (COFECUB) for the bi-national project (Grant # 774/13), coordinated by the Thermal Systems Laboratory (LST at the Pontifical Catholic University of Parana; PUCPR) and by the Conception, Optimization and Environmental Engineering Laboratory (LOCIE, at the University of Savoie Mont Blanc) that promoted the International Ph.D. School entitled "Numerical Methods for Diffusion Problems", thus motivating and enabling the writing of this book. This is an important step on the internationalization of a research line focused on the hygrothermal simulation of buildings and energy systems. The authors are most grateful to Prof. Carl-Eric Hagentoft, from the Chalmers University of Technology, for reading the book and writing the foreword. The authors also thank Prof. Carsten Rode, from the Technical University of Denmark, and Prof. Francisco Chinesta, from the École Centrale de Nantes, for their recommendations. Further, the authors express their gratitude to CAPES, for providing the Grant # 88881.067983/2014-01 to Julien Berger, and to CNPq of the Brazilian Ministry of Science, Technology and Innovation for the continuous support to LST at PUCPR.

Contents

1   A Brief History of Diffusion in Physics   1

Part I   Basics of Numerical Methods for Diffusion Phenomena
         in Building Physics

2   Heat and Mass Diffusion in Porous Building Elements   9
    2.1   A Brief Historical   9
    2.2   Heat and Mass Diffusion Models 13
    2.3   Boundary Conditions 24
    2.4   Discretization 26
    2.5   Stability Conditions 29
    2.6   Linearization of Boundary Conditions or Source Terms 30
    2.7   Numerical Algorithms 31
    2.8   Multitridiagonal-Matrix Algorithm 32
    2.9   Mathematical Model for a Room Air Domain 34
    2.10   Hygrothermal Models Used in Some Available
           Simulation Tools 36
    2.11   Final Remarks 42

3   Finite-Difference Method 45
    3.1   Numerical Methods for Time Evolution: ODE 46
           3.1.1   An Introductory Example 46
           3.1.2   Generalization 52
           3.1.3   Systems of ODEs 60
           3.1.4   Exercises 64
    3.2   Parabolic PDE 66
           3.2.1   The Heat Equation in 1D 67
           3.2.2   Nonlinear Case 75
           3.2.3   Applications in Engineering 77

| | 3.2.4 | Heat Equation in Two and Three Space Dimensions | 81 |
| | 3.2.5 | Exercises | 84 |

4 Basics in Practical Finite-Element Method 89
4.1 Heat Equation 89
    4.1.1 Weak Formulation and Test Functions 90
    4.1.2 Finite Element Representation 91
    4.1.3 Finite Element Approximation 94
4.2 Finite Element Approach Revisited 96
    4.2.1 Reference Element 96
    4.2.2 Connectivity Table 99
    4.2.3 Stiffness Matrix Construction 99
    4.2.4 Final Remarks 99

Part II Advanced Numerical Methods

5 Explicit Schemes with Improved CFL Condition 103
5.1 Some Healthy Criticism 104
5.2 Classical Numerical Schemes 105
    5.2.1 The Explicit Scheme 106
    5.2.2 The Implicit Scheme 107
    5.2.3 The Leap-Frog Scheme 108
    5.2.4 The Crank–Nicolson Scheme 109
    5.2.5 Information Propagation Speed 110
5.3 Improved Explicit Schemes 111
    5.3.1 Dufort–Frankel Method 111
    5.3.2 Saulyev Method 113
    5.3.3 Hyperbolization Method 116
5.4 Discussion 119

6 Reduced Order Methods 121
6.1 Introduction 121
    6.1.1 Physical Problem and Large Original Model 121
    6.1.2 Model Reduction Methods for Building Physics Application 122
6.2 Balanced Truncation 124
    6.2.1 Formulation of the ROM 124
    6.2.2 Marshall Truncation Method 125
    6.2.3 Building the ROM 125
    6.2.4 Synthesis of the Algorithm 126
    6.2.5 Application and Exercise 126
    6.2.6 Remarks on the Use of Balanced Truncation 129

6.3		Modal Identification	130
	6.3.1	Formulation of the ROM	130
	6.3.2	Identification Process	131
	6.3.3	Synthesis of the Algorithm	132
	6.3.4	Application and Exercise	132
	6.3.5	Some Remarks on the Use of the MIM	138
6.4		Proper Orthogonal Decomposition	139
	6.4.1	Basics	139
	6.4.2	Capturing the Main Information	140
	6.4.3	Building the POD Model	140
	6.4.4	Synthesis of the Algorithm	141
	6.4.5	Application and Exercise	141
	6.4.6	Remarks on the Use of the POD	145
6.5		Proper Generalized Decomposition	146
	6.5.1	Basics	146
	6.5.2	Iterative Solution	146
	6.5.3	Computing the Modes	147
	6.5.4	Convergence of Global Enrichment	148
	6.5.5	Synthesis of the Algorithm	148
	6.5.6	Application and Exercise	149
	6.5.7	Remarks on the Use of the PGD	151
6.6		Final Remarks	151
7		**Boundary Integral Approaches**	153
7.1		Basic BIEM	154
	7.1.1	Domain and Boundary Integral Expressions	154
	7.1.2	Green Function and Boundary Integral Formulation	156
	7.1.3	Numerical Formulation	157
7.2		Trefftz Method	159
	7.2.1	Trefftz Indirect Method	160
	7.2.2	Method of Fundamental Solutions	163
	7.2.3	Trefftz Direct Method	164
	7.2.4	Final Remarks	166
8		**Spectral Methods**	167
8.1		Introduction to Spectral Methods	167
	8.1.1	Choice of the Basis	169
	8.1.2	Determining Expansion Coefficients	176
8.2		Aliasing, Interpolation and Truncation	178
	8.2.1	Example of a Second Order Boundary Value Problem	180

8.3   Application to Heat Conduction 182
    8.3.1   An Elementary Example 182
    8.3.2   A Less Elementary Example 184
    8.3.3   A Real-Life Example 185
8.4   Indications for Further Reading 188
8.5   Appendix 1: Some Identities Involving Tchebyshev
    Polynomials 190
    8.5.1   Compositions of Tchebyshev Polynomials 193
8.6   Appendix 2: Trefftz Method 194
8.7   Appendix 3: Monte–Carlo Approach to the Diffusion
    Simulation 198
    8.7.1   Brownian Motion Generation 201
8.8   Appendix 4: An Exact Non-periodic Solution to the 1D
    Heat Equation 202
8.9   Some Popular Numerical Schemes for ODEs 204
    8.9.1   Existence and Unicity of Solutions 207

Part III   Praxis

9   Exercises and Problems 213
9.1   Discretization of Diffusion Equations 213
    9.1.1   Treatment of the Boundary Conditions 213
    9.1.2   Numerical Solution 215
9.2   Heat and Mass Diffusion: Numerical Solution 218
9.3   Whole Building Energy Simulation 222
    9.3.1   Heat Transfer 222
    9.3.2   Moisture Transfer 224
    9.3.3   Heat and Moisture Transfer 225
9.4   Heat and Mass Diffusion: Analysis of the Physical
    Behavior 225

10   Conclusions 229

References 233

Index 243

Acronyms

1D	One Dimensional
2D	Two Dimensional
3D	Three Dimensional
ADI	Alternating Direction Implicit
BE	Backward Euler
BEM	Boundary Elements Method
BER	Branch Eigenmodes Reduction
BIEM	Boundary Integral Equations Method
BTCS	Backward Time Centered Space
BVP	Boundary Value Problem
cCN	cross-Crank–Nicolson scheme
CFL	Courant-Friedrichs-Lewy
CN	Crank–Nicolson scheme
DFT	Discrete Fourier Transform
EPS	Expanded polystyrene
FCFT	Fast Cosine Fourier Transform
FDM	Finite-Difference Method
FE	Forward Euler
FEM	Finite-Element Method
FFT	Fast Fourier Transform
HAM	Heat Air Moisture
HPC	High Performance Computing
HVAC	Heating, Ventilation, and Air Conditioning
IVP	Initial Value Problem
LHS	Left Hand Side
LMM	Linear Multi-steps Methods
LOD	Locally 1D Method
LOM	Large Original Model
MIM	Modal Identification Method
MOL	Method of Lines

MTDMA	Multi-Tridiagonal-Matrix Algorithm
NA	Numerical Analysis
ODE	Ordinary Differential Equation
PDE	Partial Differential Equation
PGD	Proper Generalised Decomposition
PhD	Philosophiae Doctor
POD	Proper Orthogonal Decomposition
RHS	Right Hand Side
RK	Runge–Kutta
ROM	Reduced Order Model
RTE	Radiative Transfer Equation
SEM	Spectral Element Method
SM	Spectral Method
TDMA	Tridiagonal Matrix Algorithm
UDF	User-Defined Function
WBES	Whole-Building Energy Simulation

Chapter 1
A Brief History of Diffusion in Physics

Since the main focus of the Ph.D. school is set on the diffusion processes (molecular diffusion, heat and moisture conduction through the walls, etc.), it is desirable to explain how this research started and why the diffusion is generally modeled by *parabolic PDEs* [62]. The historic part of this chapter is partially based on [161].

First of all, let us make a general, perhaps surprising, remark, which is based on the analysis of historical investigations: understanding the microscopic world is not compulsory to propose a reliable macroscopic law. Indeed, Fourier did not know anything about the nature of heat, Ohm[1] about the nature of electricity, Fick[2] about salt solutions and Darcy[3] about the structure of porosity and water therein. In cases of the diffusion, the bridge between microscopic and macroscopic worlds was built by A. Einstein[4] in 1905. Namely, he expressed a macroscopic quantity—the diffusion coefficient—in terms of microscopic data (this result will be given below). In general, 1905 was *Annus Mirabilis*[5] for A. Einstein. During this year he published four papers in *Annalen der Physik* whose aftermath was prodigious.

The first scientific study of diffusion was performed by a Scottish chemist Thomas Graham (1805–1869). His research on diffusion was conducted between 1828 and 1833. Here we quote his first paper:

> [...] the experimental information we possess on the subject amounts to little more than the well established fact, that gases of different nature, when brought into contact, do not arrange themselves according to their density, the heaviest undermost, and the lighter uppermost, but they spontaneously diffuse, mutually and equally, through each other, and so remain in the intimate state of mixture for any length of time.

---

[1] Georg Simon Ohm (1789–1854) was a German Physicist.

[2] Adolf Eugen Fick (1829–1901) was a German physician and physiologist.

[3] Henry Darcy (1803–1858) was a French engineer in Hydraulics.

[4] Albert Einstein (1879–1955) was a German theoretical Physicist. This personality does not need to be introduced.

[5] *Annus Mirabilis* comes from Latin and stands for "extraordinary year" in English or "Wunderjahr" in German.

© Springer Nature Switzerland AG 2019
N. Mendes et al., *Numerical Methods for Diffusion Phenomena in Building Physics*,
https://doi.org/10.1007/978-3-030-31574-0_1

In 1867 James Maxwell[6] estimated the diffusion coefficient of CO_2 in the air using the results (measurements) of Graham. The resulting number was obtained within 5% accuracy. It is pretty impressive.

A. Fick[7] hold a chair of physiology in Würzburg for 31 years. His main contributions to Physics were made during a few years around 1855. When he was 26 years old he published a paper on diffusion establishing the now classical Fick's diffusion law. Fick did not realize that dissolution and diffusion processes result from the movement of separate entities of salt and water. He deduced the quantitative law proceeding in analogy with the work of Fourier [66] who modeled the heat conduction:

> […] It was quite natural to suppose that this law for diffusion of a salt in its solvent must be identical with that according to which the diffusion of heat in a conducting body takes place; upon this law Fourier founded his celebrated theory of heat, and it is the same that Ohm applied […] to the conduction of electricity […] according to this law, the transfer of salt and water occurring in a unit of time between two elements of space filled with two different solutions of the same salt, must be, *ceteris partibus*, directly proportional to the difference of concentrations, and inversely proportional to the distance of the elements from one another.

Going along this analogy, Fick assumed that the flux of matter is proportional to its concentration gradient with a proportionality factor κ, which he called "*a constant dependent upon the nature of the substances*". Actually, Fick made an error in (minus) sign which introduces the anti-diffusion and leads to an ill-posed problem. This error was not a source of difficulty for Fick since he analyzed only steady states in accordance with available experimental conditions at that time.

The fundamental article [58] devoted to the study of Brownian[8] motion and entitled "On the Motion of Small Particles Suspended in a Stationary Liquid, as Required by the Molecular Kinetic Theory of Heat" by Einstein was published on 18th July 1905 in *Annalen der Physik*. By the way, it is the most cited Einstein's paper among four works published during his *Annus Mirabilis*. This manuscript is of capital interest to us as well. Let us quote a paragraph from [58]:

> In this paper it will be shown that, according to the molecular kinetic theory of heat, bodies of a microscopically visible size suspended in liquids must, as a result of thermal molecular motions, perform motions of such magnitudes that they can be easily observed with a microscope. It is possible that the motions to be discussed here are identical with so-called Brownian molecular motion; however, the data available to me on the latter are so imprecise that I could not form a judgment on the question […]

Einstein was the first to understand that the main quantity of interest is the mean square displacement $\langle X^2(t) \rangle$ and not the average velocity of particles. Taking into account the discontinuous nature of particle trajectories, the velocity is meaningless.

---

[6]James Clerk Maxwell (1831–1879) was a Scottish Physicist, famous for Maxwell equations.

[7]Adolf Fick was the author of the first treatise of *Die medizinische Physik* (Medical Physics) (1856) where he discussed biophysical problems, such as the mixing of air in the lungs, the work of the heart, the heat economy of the human body, the mechanics of muscular contraction, the hydrodynamics of blood circulation, etc.

[8]Robert Brown (1773–1858) was a Scottish Botanist.

Before publication of [58] atoms in Physics were considered as a useful, but purely theoretical concept. Their reality was seriously debated. W. Ostwald,[9] one of the leaders of the anti-atom school, later told A. Sommerfeld[10] that he had been convinced of the existence of atoms by Einstein's complete explanation of Brownian motion.

Without knowing it, Einstein in answered a question posed the same year by K. Pearson[11] in a Letter published in Nature [156]:

> Can any of your readers refer me to a work wherein I should find a solution of the following problem, or failing the knowledge of any existing solution provide me with an original one? I should be extremely grateful for aid in the matter.
>
> A man starts from a point O and walks ℓ yards in a straight line; he then turns through any angle whatever and walks another ℓ yards in a second straight line. He repeats this process n times. I require the probability that after these n stretches he is at a distance between r and $r + dr$ from his starting point, O.
>
> The problem is one of considerable interest, but I have only succeeded in obtaining an integrated solution for two stretches. I think, however, that a solution ought to be found, if only in the form of a series in powers of $\frac{1}{n}$, when n is large.

The expression "*random walk*" was probably coined in Pearson's Letter [156].

Let us follow Einstein's study of the Brownian motion. Consider a long thin tube filled with water (it allows us to consider only one spatial dimension). At the initial time $t = 0$ we inject a unit amount of ink at $x = 0$. Let $u(x, t)$ denote the density of ink particles in location $x \in \mathbb{R}$ and at time $t \geqslant 0$. So, initially we have

$$u(x, 0) = \delta(x), \tag{1.1}$$

where $\delta(x)$ is the Dirac distribution centered at zero.

Then, suppose that the probability density of the event that an ink particle moves from x to $x + \Delta x$ in a small time Δt is $\rho(\Delta x, \Delta t)$. Then, we have

$$u(x, t + \Delta t) = \int_{-\infty}^{+\infty} u(x - \Delta x, t)\, \rho(\Delta x, \Delta t)\, \mathrm{d}(\Delta x).$$

Assuming that solution $u(x, t)$ is smooth, we can apply the Taylor[12] formula:

$$u(x, t + \Delta t) = \int_{-\infty}^{+\infty} \left(u - u_x \cdot \Delta x + \frac{1}{2} u_{xx} \cdot (\Delta x)^2 + \cdots \right) \rho(\Delta x, \Delta t)\, \mathrm{d}(\Delta x). \tag{1.2}$$

---

[9]Friedrich Wilhelm Ostwald (1853–1932) was a Latvian Chemist. He received a Nobel prize in Chemistry in 1909 for his works on catalysis.

[10]Arnold Johannes Wilhelm Sommerfeld (1868–1951) was a German theoretical Physicist. He served as the Ph.D. advisor for several Nobel prize winners.

[11]Karl Pearson (1857–1936) was an English Statistician.

[12]Brook Taylor (1685–1731) was an English Mathematician.

Now let us recall that ρ is a Probability Density Function (PDF). Thus,

Normalisation $\int_{-\infty}^{+\infty} \rho(\Delta x, \Delta t) \, \mathrm{d}(\Delta x) = 1$

Symmetry   we can assume that the diffusion process is symmetric in space, i.e.

$$\rho(\Delta x, \Delta t) = \rho(-\Delta x, \Delta t), \qquad \forall \Delta x, \ \Delta t \geqslant 0.$$

Consequently,

$$\int_{-\infty}^{+\infty} \Delta x \, \rho(\Delta x, \Delta t) \, \mathrm{d}(\Delta x) = 0.$$

Variance   We can also assume that the variance if finite and Einstein assumed additionally that the variance is *linear* in Δt:

$$\int_{-\infty}^{+\infty} (\Delta x)^2 \, \rho(\Delta x, \Delta t) \, \mathrm{d}(\Delta x) = \mathscr{D} \Delta t,$$

where $\mathscr{D} > 0$ is the so-called *diffusion constant*.

By incorporating these results into Eq. (1.2) and rearranging the terms, it becomes

$$\frac{u(x, t + \Delta t) - u(x, t)}{\Delta t} = \frac{1}{2} \mathscr{D} u_{xx}(x, t) + \cdots .$$

Taking the limit $\Delta t \to 0$ we obtain straightforwardly

$$u_t = \frac{1}{2} \mathscr{D} u_{xx} . \tag{1.3}$$

The Initial Value Problem (IVP) (1.1) for this linear parabolic equation (1.3) can be solved exactly:

$$u(x, t) = \frac{1}{\sqrt{2\pi \mathscr{D} t}} \, \mathrm{e}^{-\frac{x^2}{2\mathscr{D} t}} .$$

The last solution is also known as the Green function.

However, the main result of Einstein's paper [58] is the following formula:

$$\mathscr{D} = \frac{RT}{f N_a}, \tag{1.4}$$

where

R the ideal gas constant, i.e. $R \approx 8.3144598 \, \frac{\mathrm{J}}{\mathrm{K \cdot mol}}$

T the absolute temperature

f the friction coefficient

N_a Avogadro's[13] number, i.e. $N_a \approx 6.022140857 \, \mathrm{mol}^{-1}$

---

[13]Lorenzo Romano Amedeo Carlo Avogadro di Quaregna e di Cerreto (1776–1856) was an Italian scientist.

Formula (1.4) along with the observation of the Brownian motion enabled J. Perrin[14] to produce the first historical estimation of Avogadro's constant $N_a \approx 6\,\text{mol}^{-1}$.

Exercise 1 Read a fascinating paper by A. Einstein on the Hydrodynamics of tea leaves [59]. By the way, the Author attests that Tea is very stimulating for intellectual activities. ☺

---

[14] Jean Baptiste Perrin (1870–1942) was a French Physicist who was honoured with the Nobel prize in Physics in 1926 for the confirmation of the atomic nature of matter (through the observation of Brownian motion).

Part I
Basics of Numerical Methods for Diffusion Phenomena in Building Physics

Chapter 2
Heat and Mass Diffusion in Porous Building Elements

In the field of building physics, diffusion phenomena started to be extensively modeled in the 70s (because of the oil crisis) to develop building-performance-simulation programs for the adoption of rational policies of energy conservation. However, existing tools might still present inconsistent scenarios of the actual occurrences in buildings, especially in the heat and mass transfer domain. The mathematical description for predicting building hygrothermal dynamics is complex because of the nonlinearities, three-dimensionality of transport phenomena, and interdependence of several variables. The parametric uncertainties in modeling, simulation time steps, external climate, building schedules, hygrothermal properties, ground temperature and moisture storage, and transport also contribute to the increase in this complexity.

Moreover, the moisture in porous building elements, such as furniture and envelope, implies an additional mechanism of transport adsorbing or releasing of latent heat of vaporization, affecting the hygrothermal building performance or even causing mold growth. To solve heat and mass diffusion phenomena in building physics, models can be divided into three domains: building elements (walls, roof, and furniture), ground, and room air.

This chapter first presents a brief history of the diffusion in building elements. Next, mathematical models for predicting heat and moisture diffusion through porous building elements are provided along with some numerical techniques, a room air lumped model, some hygrothermal models available in simulation tools, and final remarks.

2.1 A Brief Historical

If the mass transfer is disregarded, we can assess the thermal performance of the opaque envelope considering only the Fourier's equation in a nonsteady regime as

© Springer Nature Switzerland AG 2019
N. Mendes et al., *Numerical Methods for Diffusion Phenomena in Building Physics*,
https://doi.org/10.1007/978-3-030-31574-0_2

$$\rho_0 \, C_m(T) \frac{\partial T}{\partial t} \; = \; \nabla \cdot (\lambda(T) \, \nabla T) \; , \tag{2.1}$$

where ρ_0 is the material density (kg/m^3), C_m is the mean specific heat (J/kg K), T is the temperature ($^\circ$C), and λ is the thermal conductivity (W/m K).

This purely conductive heat model has been simplified using constant thermophysical properties and solved in building-energy-simulation programs, such as DOE-2, BLAST, and TRNSYS, since the 70s using the response factor method [1], as illustrated in Fig. 2.1. However, it does not allow the consideration of variable properties, source terms, and risk of interstitial condensation. If moisture is considered, this purely conductive heat model should be extended to include the moisture balance as well.

To predict the risk of interstitial moisture condensation, the Glaser method (DIN 4108) [81], based on graphical methods for investigation of diffusion processes, has been extensively used. Although the steady-state diffusion-based method might provide good results for risk assessment and design of moisture-safe constructions, especially for light weight ones, model limitations such as the disregard of hygroscopic sorption, liquid transport, and transient effects prevent the method from simulating and providing accurate and realistic information about hygrothermal performance of building envelopes [118]. However, over the last 30 years, with significant progress in computer hardware, the Glaser method has inspired the development of detailed heat, air, and moisture (HAM) mathematical models and simulation tools for the analysis of a single porous building element or the whole building. The models have been progressively improved so that different models are available, and a very brief and non-complete overview is presented as follows.

Silveira-Neto proposed analytical solutions [182] to solve monolithic porous walls based on the model of Philip and DeVries [162], assuming constant properties and different boundary conditions.

Cunningham developed a mathematical model for hygroscopic materials in flat structures; it uses an electrical analogy with resistances for vapor flow and an exponential approximation function with constant mass transport coefficients [45–48]. Kerestecioglu and Gu investigated the diffusion phenomenon by using evaporation–condensation theory in the pendular state (unsaturated-liquid-flow stage) [113]. The application of this theory is limited to low moisture content.

Pedersen presented a transient model for analyzing the hygrothermal behavior of building constructions [157] by considering both vapor and liquid transfer, and a hysteresis model, which allows to consider the scanning curves in an adsorption/desorption process instead of a mean curve, as is normally used. His model, which was developed at the Danish Technical University (DTU)—originated the MATCH program, which was apparently the first simulation tool available for predicting heat and moisture performance of porous building elements. The MATCH model was included in the Danish building simulation program BSim [167], allowing the prediction of vapor pressure distribution in construction elements when the whole building is simulated.

Burch and Thomas investigated moisture accumulation in wooden-frame walls subjected to winter [26]. Their model originated the computer program MOIST [27] by using the finite-difference method to estimate the heat and mass transfer through composite walls under nonisothermal conditions. In addition, the model neglects the liquid transport through porous building elements. El Diasty et al. used an analytical approach that assumed isothermal conditions and constant transport coefficients for an unsteady model of moisture assessment in buildings [60]. Liesen [130] used an evaporation–condensation theory and a response factor method to develop and implement a model of heat and mass transfers in the building thermal simulation program IBLAST (Integrated Building Loads Analysis and System Thermodynamics). In this, the hygrothermal property variations were neglected and liquid transfer was not considered. Yik et al. [205] developed a fast model integrated with air-conditioning system component models that employs the evaporation–condensation theory with differential permeabilities.

Bueno et al. studied the heat and mass transfer phenomena through roofs [25] using a finite-difference base model, and concluded that condensation (or rain) on the roof at night, followed by evaporation in the morning, may be very important to reach naturally comfortable conditions in hot and humid climates. However, yearly simulations were not executed at that time as he had to use a 0.1 s time step because of numerical divergence problems. Künzel describes in details the formulation of the simultaneous prediction of heat and moisture transport in building components in one and two dimensions, which originated the widely used WUFI program developed at the Fraunhofer Institute of Building Physics [119]. WUFI was one of the first simulation tools to predict 2-D heat and moisture transfer in building components [118].

Mendes et al. observed that moisture effects are extremely significant during the calculation of thermal loads. Especially, when the air conditioner is turned on in the early morning, the latent heat flux rises substantially due the low relative humidity of the room air imposed by the machine [137, 138, 142].

The Mendes model gave rise to the UMIDUS PC simulation tool [137, 141] for the prediction of temperature and moisture content from basic properties, such as sorption isotherm, mercury porosimetry, and dry-basis thermal conductivity. This model and its improvements have been devised for the whole-building simulation program Domus, since 2001, at the Pontifical Catholic University of Parana, Brazil. Santos et al. presented the SOLUM program [55] for 3D prediction of temperature and moisture content in the ground by using an extended 3D version of the model presented in UMIDUS. Grunewald presented a mathematical model for the development of a simulation tool at the Technological University of Dresden [85]. This model has been improved since 1989 (named DIM1 at that time), and an extension to the coupled heat and moisture transport was released and was renamed as DIM3 [86]. This software has been continuously updated, and with a graphical user interface, the software was renamed as the well-known DELPHIN program; the current version is called DELPHIN 5. Details about this model can be found in [10]. Karagiozis presented a hygrothermal modeling of building materials by using a model called MOISTURE-EXPERT [110], which included different levels of unintentional water penetration, for the evaluation of hygrothermal performance. In addition, he investigated the performance of the model for a conventional stucco-clad wooden-frame wall. Hagentoft et al. presented a very comprehensive assessment method in numerical prediction models for combined heat, air, and moisture transfer in building components [91]. In addition, they presented models and performed benchmarks for 1D cases by using different models developed in Europe. At the Chalmers University of Technology, Sweden, Kalagasidis developed a modular-based environment named HAM-Tools [108] by using the graphical programming language Simulink, which is an integrated simulation tool for diffusion phenomena in building physics. Tariku et al. combined a building-envelope model and an indoor model [187], and coupled them to form a whole-building hygrothermal model, called HAMFitPlus. All the aforementioned models facilitated in the development of many simulation tools to predict the heat and moisture transfer through porous building walls by mostly using calculation methods that are conditionally stable.

In the late 80s, air transport was also considered in the coupled formulation of heat and moisture through porous building envelopes, especially in cold countries where high insulation materials are employed. Since the 90s, salt diffusion has also been a factor of great concern; however, some difficulties, such as chemical reactions and the correct definition of boundary conditions, limit the enhancement progress of the formulation.

Models have nearly the same origin. The main differences among them are related to the particular assumptions used and the choice of driving potentials. Some of the HAM diffusive transfer tools have been developed in the last 25 years and some are commented upon in detail in the IEA Annex 24 (1996); these include 1D-HAM, WUFI, MATCH, HYGRAN24, JOKE, and LATENITE. During the IEA Annex 41 project, from 2004–2007, some HAM models were presented and integrated with whole-building simulation models, such as DELPHIN, MATCH in BSim, WUFI in WUFI+, and UMIDUS in DOMUS. Woloszyn and Rode detailed tools tested during the IEA Annex 41 [200] that were mainly developed from 1D building element to the whole-building. Whole-Building Energy Simulation (WBES) tools, such as TRNSYS, ESP, and EnergyPlus, developed and widely used since the 70s also have combined heat and moisture diffusion models in their codes. The Wufi model was also implemented in EnergyPlus, allowing a more accurate prediction of heat and moisture content through building elements [44, 82]. In general, diffusion models are based on partial differential equations (PDEs) solved by using traditional methods, such as finite elements, finite differences, and finite volumes, applied to multilayered components of the building envelope. Normally, codes do not always have an algorithm with a robust solver to simultaneously calculate the temperature and moisture content distributions. Instead, they have stability criteria and time step control devices to reduce numerical convergence problems. Nevertheless, in the last few years, new heat and moisture diffusion models to be applied to building physics have been developed to solve multidimensional diffusion processes. These include the models presented in [136, 178, 184, 192] that can be used in 3D complex geometry configurations by using powerful commercial solvers such as COMSOL and ANSYS-CFD (CFX and FLUENT). However, the use of these tools for whole-building simulation is extremely time consuming, thus requiring considerable research to make them usable by building engineers, architects, and consultants. To solve this issue of computational costs, researchers have recently attempted to use model reduction methods to solve heat and moisture transfer. For example, the coupling of a Proper-Generalized-Decomposition (PGD) Reduced Order Model (ROM) with the whole-building simulation program DOMUS presented in [15–17].

Figure 2.1 illustrates the evolution of some of the heat and mass diffusion models and tools for building energy simulation. The development of software in this field is correlated to the world energy and environment crises and to the funding for International Energy Agency projects. Some ongoing research and development for the next years seem to be focused on the use of advanced numerical methods and the investigation of 3D heat and moisture transfer combined with CFD. Indeed, owing to the dramatic progress in computer hardware and increase in urban flood and heat-island related problems, research is also being undertaken on the integration of heat and moisture models at the district scale. In addition, approaches at the nano/micropore scale (Nano-HAM), which are based on the construction of kinetic models for studying the processes of moisture migration, are very promising in the future. They might enable to obtain, for instance, precise hygrothermal transport coefficients and sorption isotherm curves that are used in macroscopic diffusive and convective HAM models from microstructure information.

2.2   Heat and Mass Diffusion Models

Strongly coupled diffusion equations are found in the combined heat and mass transfer problems through porous materials in building physics. Figure 2.2 schematically shows the moisture transport process through the porous microstructure under the presence of a vapor concentration gradient. Porous building materials, such as lime mortar and brick, have a large distribution of pore diameters. Small pores can be easily filled up with water (dark region), and the transport occurs predominantly in this region through vapor condensation followed by capillary migration and evaporation at the opposite end of the porous cavity. In other pores with greater diameters, the moisture transport occurs mainly through vapor diffusion. At very high values of relative humidity, capillary transport becomes more relevant and the moisture transport is considerably increased. The higher the moisture transport, the higher are the gradients and coupling between the governing equations.

Moisture is stored through physical adsorption process, and in smaller pores through capillary condensation. The thickness of the adsorbed water layer increases with relative humidity. After snap-off, the pore cavity is quickly filled up with water and the capillary condensation mechanism starts.

The moisture diffusion phenomenon may play a very important role on the whole heat transfer process through building elements, as phase change is present and the physical properties are highly dependent on temperature and moisture content. For

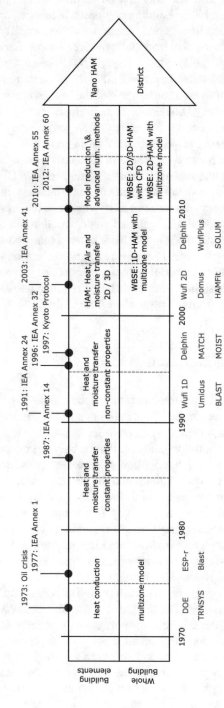

Fig. 2.1 Evolution of the development of the heat and mass diffusion models for building energy simulation tools

Fig. 2.2 Moisture transport through a porous material [138]

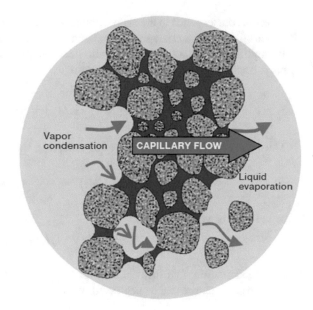

instance, thermal conductivities may vary two or even three times from the dry to the fully water-saturated state.

To obtain the governing formulation, mass- and energy-balance conservation equations must be applied into an elemental control volume.

Mass conservation : For the liquid phase, we can write the mass balance as

$$\frac{\partial}{\partial t}(\rho_0 \omega_l) = -\nabla \cdot \mathbf{j}_l + \sigma_l, \qquad (2.2)$$

where ω_l corresponds to the liquid moisture content in terms of mass (kg of water/kg of dry material) and $\mathbf{j}_l$ represents the (liquid) water flow. By multiplying Eq. (2.2) with the ratio between the (liquid) water density and dry-basis material density (ρ_l/ρ_0), we obtain:

$$\frac{\partial \theta_l}{\partial t} = -\nabla \cdot (\mathbf{j}_l/\rho_l) - \frac{\rho_l}{\rho_0}\sigma_l, \qquad (2.3)$$

where $\frac{\rho_l}{\rho_0}\sigma_l$ (s^{-1}) represents the source of vapor or liquid due to phase change and θ_l is the liquid volumetric water moisture content (m^3 of water per m^3 of material). For the vapor phase, we obtain

$$\frac{\partial}{\partial t}(\rho_0 \omega_v) = -\nabla \cdot \mathbf{j}_v + \sigma_v, \qquad (2.4)$$

where ω_v corresponds to the vapor moisture content in terms of mass (kg of water vapor/kg of dry material) and $\mathbf{j}_v$ represents the water vapor flow.

By multiplying Eq. (2.4) with the ratio between the water vapor density and dry-basis material density (ρ_v/ρ_0), we obtain:

$$\frac{\partial \theta_v}{\partial t} = -\nabla \cdot (\mathbf{j}_v/\rho_v) + \frac{\rho_v}{\rho_0}\sigma_v,$$

Thus, for the moisture content (vapor + liquid), the mass balance can be written as

$$\frac{\partial}{\partial t}\left[\rho_0\left(\omega_l + \omega_v\right)\right] = -\nabla \cdot (\mathbf{j}_l + \mathbf{j}_v)$$

or in terms of volume as

$$\frac{\partial \theta}{\partial t} = -\nabla \cdot (\mathbf{j}/\rho_l) . \tag{2.5}$$

Energy conservation: If the heat transfer through radiation and convection within the porous structure can be neglected, we can write

$$\frac{\partial}{\partial t}\left[\rho_0\left(h_0 + h_l\,\omega_l + h_v\,\omega_v\right)\right] = -\nabla \cdot (\mathbf{q} + h_l\,\mathbf{j}_l + h_v\,\mathbf{j}_v) ,$$

where $\mathbf{q}$ represents the heat flux given by the Fourier's law and h_s, h_0, and h_v are the enthalpies for the solid, liquid, and vapor phases, respectively.

The porosity η can be written as the sum of liquid and vapor volumetric moisture contents:

$$\eta = \theta_l + \theta_v,$$

where the volume occupied by moist air is equal to the volume occupied by vapor $(\theta_{\mathrm{air}} = \theta_v)$. Furthermore, the vapor mass content can similarly be written in terms of the liquid mass content as

$$\omega_v = \left(\eta - \omega_l\frac{\rho_0}{\rho_l}\right)\frac{\rho_v}{\rho_0}.$$

By considering the water vapor as a perfect gas, we obtain

$$\omega_v = \left(\eta - \omega_l\frac{\rho_0}{\rho_l}\right)\frac{P_v(\Psi, T)\,M}{RT\rho_0}. \tag{2.6}$$

The vapor pressure in Eq. (2.6) is a function of the water potential Ψ and temperature. M and R are the molar mass (kg/mol) and the universal gas constant (J/K/mol), respectively.

The Kelvin's law establishes that when there is thermodynamic equilibrium between the liquid and vapor phases, they are at the same energy level, i.e.

$$P_v = P_s(T) \exp\left(\frac{\Psi M g}{R T}\right) = P_s(T) \phi.$$

Then, we obtain the following expression:

$$\omega_v = -\omega_l \frac{M P_s(T) \phi}{R T \rho_l} + \eta \frac{M P_s(T) \phi}{R T \rho_0}, \tag{2.7}$$

where ϕ is the relative humidity of moist air and P_s is the saturated water pressure. Thus, the derivative of Eq. (2.7) can be written as

$$\frac{\partial \omega_v}{\partial t} = -A \frac{\partial \omega_l}{\partial t} + B \frac{\partial \Psi}{\partial t} + C \frac{\partial T}{\partial t}, \tag{2.8}$$

where

$$A = \frac{M P_s(T) \phi}{R T \rho_l}, \tag{2.9a}$$

$$B = \Theta_v g P_s \left(\frac{M}{R T}\right)^2 \frac{\phi}{\rho_0}, \tag{2.9b}$$

$$C = \frac{\Theta_v M P_s}{R T} \frac{\phi}{\rho_0} \left[\frac{1}{P_s} \frac{d P_s}{d T} - \frac{\Psi M g}{(R T)^2} - \frac{1}{T}\right]. \tag{2.9c}$$

Therefore, we can write the mass conservation balance as

$$(1 - A) \frac{\partial \omega_l}{\partial t} + B \frac{\partial \Psi}{\partial t} + C \frac{\partial T}{\partial t} = -\frac{1}{\rho_0} \nabla \cdot (\mathbf{j})$$

or

$$(1 - A) \frac{\partial \omega_l}{\partial t} + B \frac{\partial \Psi}{\partial t} + C \frac{\partial T}{\partial t} = -\frac{1}{\rho_0} \nabla \cdot (\mathbf{j}_l + \mathbf{j}_v),$$

As such, the energy balance equation can be written as

$$(h_l - A h_v) \frac{\partial \omega_l}{\partial t} + B h_v \frac{\partial \Psi}{\partial t} + (c_m + h_v C) \frac{\partial T}{\partial t} = -\frac{1}{\rho_0} \cdot \nabla (\mathbf{q} + h_l \mathbf{j}_l + h_v \mathbf{j}_v),$$

where c_m represents the mean specific heat given by

$$c_m = c_0 + \omega_l c_l + \omega_v c_v,$$

where the three right-hand terms represent the specific heat of solid, liquid, and vapor phases.

From thermodynamics, we know that the specific heat for a perfect gas can be given as

$$c = \frac{\partial h}{\partial t},$$

which allows us to obtain the following by integrating with the liquid phase:

$$h_l - h_l^r = c_l \, (T - T_r) \ .$$

By adopting, for convenience, a null reference liquid enthalpy – h_l^r, we obtain the following expression for the liquid phase. For enthalpy,

$$h_l = c_l \, (T - T_r) \ .$$

By integrating with the vapor phase, we obtain

$$h_v - h_v^r = c_v \, (T - T_r) \ ,$$

where the reference vapor enthalpy—h_v^r—is considered equal to the latent heat of vaporization at the reference temperature, which allows us to write

$$h_v = \mathscr{L}(T_r) + c_v \, (T - T_r) \ .$$

Then, we can define the latent heat of vaporization $\mathscr{L}$ as

$$\mathscr{L} = h_v - h_l = \mathscr{L}(T_r) + c_v \, (T - T_r) - c_l \, (T - T_r) \ .$$

This allows us to rewrite the energy governing equation as

$$(h_l - A \, h_v) \, \frac{\partial \omega_l}{\partial t} + B \, h_v \, \frac{\partial \Psi}{\partial t} + (c_m + h_v \, C) \, \frac{\partial T}{\partial t} =$$
$$- \frac{1}{\rho_0} \nabla \, (\mathbf{q} + c_l \, (T - T_r) \, \mathbf{j}_l + (\mathscr{L}(T_r) + c_v \, (T - T_r)) \, \mathbf{j}_v)$$

or as

$$(h_l - A \, h_v) \, \frac{\partial \omega_l}{\partial t} + B \, h_v \, \frac{\partial \Psi}{\partial t} + (c_m + h_v \, C) \, \frac{\partial T}{\partial t} = -\frac{1}{\rho_0} \nabla \cdot \mathbf{q} -$$
$$\frac{1}{\rho_0} \, (c_l \, (T - T_r) \, \nabla \cdot \mathbf{j}_l + (\mathscr{L}(T_r) + c_v \, (T - T_r)) \, \nabla \cdot \mathbf{j}_v + c_l \mathbf{j}_l \nabla T + c_v \mathbf{j}_v \nabla T) \ .$$

If the liquid and vapor moisture flows are replaced by their transient terms, we obtain

$$(h_l - A\,h_v)\,\frac{\partial \omega_l}{\partial t} + B\,h_v\,\frac{\partial \Psi}{\partial t} + (c_m + h_v\,C)\,\frac{\partial T}{\partial t} =$$
$$-\frac{1}{\rho_0}\nabla \cdot \mathbf{q} - \frac{1}{\rho_0}\left(c_l\,(T - T_r)\left[-\rho_0\frac{\partial \omega_l}{\partial t} + \sigma_l\right]\right.$$
$$\left. + (\mathcal{L}(T_r) + c_v\,(T - T_r))\left[-\rho_0\frac{\partial \omega_v}{\partial t} + \sigma_v\right] + c_l\mathbf{j}_l\nabla T + c_v\mathbf{j}_v\nabla T\right), \tag{2.10}$$

where σ_l and σ_v respectively represent the vapor and liquid sources due to phase change

$$\sigma_v = -\sigma_l = \sigma. \tag{2.11}$$

By substituting Eqs. (2.8) and (2.11) into Eq. (2.10), we obtain the following expression for energy balance:

$$(h_l - A\,h_v)\,\frac{\partial \omega_l}{\partial t} + B\,h_v\,\frac{\partial \Psi}{\partial t} + (c_m + h_v\,C)\,\frac{\partial T}{\partial t}$$
$$= -\frac{1}{\rho_0}\nabla \cdot \mathbf{q} - \frac{1}{\rho_0}\left(c_l\,(T - T_r)\left[-\rho_0\frac{\partial \omega_l}{\partial t} - \sigma\right]\right.$$
$$\left. + (\mathcal{L}(T_r) + c_v\,(T - T_r))\left[\rho_0\left(A\frac{\partial \omega_v}{\partial t} - B\frac{\partial \Psi}{\partial t} - C\frac{\partial T}{\partial t}\right) + \sigma\right]\right.$$
$$\left. + c_l\mathbf{j}_l\nabla T + c_v\mathbf{j}_v\nabla T\right),$$

or

$$(h_l - A\,h_v)\,\frac{\partial \omega_l}{\partial t} + B\,h_v\,\frac{\partial \Psi}{\partial t} + (c_m + h_v\,C)\,\frac{\partial T}{\partial t}$$
$$= -\frac{1}{\rho_0}\nabla \cdot \mathbf{q} - \frac{1}{\rho_0}\left\{c_l\,(T - T_r)\left[-\rho_0\frac{\partial \omega_l}{\partial t}\right]\right.$$
$$\left. + (\mathcal{L}(T_r) + c_v\,(T - T_r))\left[\rho_0\left(A\frac{\partial \omega_v}{\partial t} - B\frac{\partial \Psi}{\partial t} - C\frac{\partial T}{\partial t}\right)\right]\right.$$
$$\left. + c_l\mathbf{j}_l\nabla T + c_v\mathbf{j}_v\nabla T + \sigma\left[\mathcal{L}(T_r) + c_v\,(T - T_r) - c_l\,(T - T_r)\right]\right\}.$$

This can give

$$(h_l - A\,h_v)\,\frac{\partial \omega_l}{\partial t} + B\,h_v\,\frac{\partial \Psi}{\partial t} + (c_m + h_v\,C)\,\frac{\partial T}{\partial t} - c_l\,(T - T_r)\,\frac{\partial \omega_v}{\partial t}$$
$$- (\mathcal{L}(T_r) + c_v\,(T - T_r))\left[-A\frac{\partial \omega_v}{\partial t} + B\frac{\partial \Psi}{\partial t} + C\frac{\partial T}{\partial t}\right]$$
$$= -\frac{1}{\rho_0}\nabla \cdot \mathbf{q} - \frac{1}{\rho_0}\{c_l\mathbf{j}_l\nabla T + c_v\mathbf{j}_v\nabla T + \mathcal{L}\sigma\}$$

or

$$c_m \frac{\partial T}{\partial t} = -\frac{1}{\rho_0} \nabla \cdot \mathbf{q} - \frac{1}{\rho_0} \{c_l \mathbf{j}_l \nabla T + c_v \mathbf{j}_v \nabla T + \mathscr{L}\sigma\}.$$

However,

$$\sigma = \sigma_v = \frac{\partial \left(\rho_0 \omega_v\right)}{\partial T} + \nabla \cdot (\mathbf{j}_v)$$

and

$$\sigma = \rho_0 \left[-A\frac{\partial \omega_v}{\partial t} + B\frac{\partial \Psi}{\partial t} + C\frac{\partial T}{\partial t}\right] + \nabla \cdot (\mathbf{j}_v).$$

This allows us to write the energy conservation equation as

$$(c_m + \mathscr{L}C)\frac{\partial T}{\partial t} + \mathscr{L}\left(-A\frac{\partial \omega_v}{\partial t} + B\frac{\partial \Psi}{\partial t}\right) =$$
$$-\frac{1}{\rho_0}\nabla \cdot \mathbf{q} - \frac{1}{\rho_0}(c_l \mathbf{j}_l \nabla T + c_v \mathbf{j}_v \nabla T + \mathscr{L}\nabla \mathbf{j}_v).$$

Nevertheless, by considering that the vapor mass content (ω_v) is much lower than that of the liquid (ω_l) and by neglecting the sensible heat transfer in the liquid ($c_l \mathbf{j}_l \nabla T$) and gaseous ($c_v \mathbf{j}_v \nabla T$) phases, we can write the energy governing equation comprehensively as

$$\rho_0 c_m \frac{\partial T}{\partial t} = -\nabla \cdot \mathbf{q} - \mathscr{L}\nabla \cdot \mathbf{j}_v. \tag{2.12}$$

Therefore, the heat and moisture governing equations are given by Eqs. (2.5) and (2.12). However, they are still written in terms of vapor and liquid flows ($\mathbf{j}_l + \mathbf{j}_v$), which are represented in literature in different forms.

A well-known heat and moisture model is the model of Philip and De Vries [162], which was developed for soils based on temperature and moisture content gradients as the driving potentials for heat and moisture fluxes. In addition, this model comprises the water liquid transfer, which is important to be considered, especially in humid climates. This model is in accordance with the governing equations presented earlier (Eqs. (2.5) and (2.12)) and has been implemented in Umidus software. This software is developed to model coupled heat and moisture transfer within porous building elements by avoiding limitations such as low moisture content. The development of Umidus is presented in [141], and it was also implemented and extended in the Domus whole-building simulation program.

From a macroscopic point of view, the partial differential governing equations, considering the Fourier's law and variable properties, give

$$\rho_0 c_m(T, \theta) \frac{\partial T}{\partial t} = \nabla \cdot (\lambda(T, \theta) \nabla T) - \mathscr{L}(T) \nabla \cdot \mathbf{j}_v \qquad (2.13a)$$

$$\frac{\partial \theta}{\partial t} = -\nabla \cdot \left(\frac{\mathbf{j}}{\rho_l} \right), \qquad (2.13b)$$

where ρ_0 is the solid matrix density (kg/m^3), c_m is the mean specific heat (J/kg K), T is the temperature ($^\circ$C), λ is the thermal conductivity (W/m K), $\mathscr{L}$ is the latent heat of vaporization (J/kg), θ is the volumetric moisture content (m^3/m^3), $\mathbf{j}_v$ is the vapor flow (kg/m^{2}s), $\mathbf{j}$ is the total flow (kg/m^{2}s), and ρ_l is the water density (kg/m^3). Note that Eq. (2.13a) differs from Fourier's equation for transient heat flow, Eq. (2.1), by an added convective transport term (because of moisture diffusion associated with evaporation and condensation of water in the pores of the medium) and dependence on the moisture content (it is coupled to Eq. (2.3)). The driving forces for heat, liquid, and vapor transfer in this Philip and De Vries model are temperature and moisture gradients.

The total $1D$ moisture flow ($\mathbf{j}$), given by adding the vapor flow ($\mathbf{j}_v$) and liquid flow ($\mathbf{j}_l$), can be described using the Philip and De Vries model [162] as

$$\left(\frac{\mathbf{j}}{\rho_l} \right) = -D_T(T, \theta) \frac{\partial T}{\partial x} - D_\theta(T, \theta) \frac{\partial \theta}{\partial x},$$

where $D_T = D_{T_l} + D_{T_v}$ and $D_\theta = D_{\theta_l} + D_{\theta_v}$. Here D_{T_l} is the liquid-phase transport coefficient associated to a temperature gradient (m^2/s K), D_{T_v} is the vapor-phase transport coefficient associated to a temperature gradient (m/s K), D_{θ_l} is the liquid-phase transport coefficient associated to a moisture content gradient (m^2/s), D_{θ_v} is the vapor-phase transport coefficient associated to a moisture content gradient (m^2/s), D_T is the mass transport coefficient associated to a temperature gradient (m^2/s K), and D_θ is the mass transport coefficient associated to a moisture content gradient (m^2/s). A 3D formulation based on this model was implemented in the Solum program and is described in [55]. The total $3D$ moisture flow ($\mathbf{j}$), given by adding the vapor flow ($\mathbf{j}_v$) and liquid flow ($\mathbf{j}_l$), is described as

$$\frac{\mathbf{j}}{\rho_l} = -\left(D_T(T, \theta) \frac{\partial T}{\partial x} - D_\theta(T, \theta) \frac{\partial \theta}{\partial x} \right) \mathbf{e_i}$$
$$- \left(D_T(T, \theta) \frac{\partial T}{\partial y} - D_\theta(T, \theta) \frac{\partial \theta}{\partial y} \right) \mathbf{e_j}$$
$$- \left(D_T(T, \theta) \frac{\partial T}{\partial z} - D_\theta(T, \theta) \frac{\partial \theta}{\partial z} + K_l \right) \mathbf{e_k},$$

where K_l is the hydraulic conductivity that enables to calculate the liquid transport due to the gravity effect. This effect is negligible for the vapor phase.

This Philip and De Vries-based model, for the modeling of heat and moisture transfer through porous building elements, uses the moisture content gradient as a driving potential. However, it is well-known that there is discontinuity in the moisture content profile at the interface between two porous materials because of their different hygroscopic capacities caused by their unequal pore size distribution functions; a material with high hygroscopicity retains more moisture at the same relative humidity and temperature.

To model porous building envelopes made up of different porous materials, with the runabout of moisture content discontinuity problem, Mendes and Philippi [139] presented a moisture-content based mathematical model that considers the discontinuity effects at the interface between two dissimilar porous materials. They proposed the following additional equation to be implemented at the interface between materials A and B:

$$\left(\frac{\partial \phi}{\partial \theta}\right)_A^{\mathrm{prev}} \left(\theta_A(s) - \theta_A^{\mathrm{prev}}(s)\right) = \left(\frac{\partial \phi}{\partial \theta}\right)_B^{\mathrm{prev}} \left(\theta_B(s) \dot{-} \theta_B^{\mathrm{prev}}(s)\right),$$

where θ_A and θ_B are the moisture contents at each side of the interface.

In many circumstances, the direct use of moisture content in some mathematical models and computer codes can be very appropriate as it can be more computationally viable. In addition, most of time, moisture content is a more useful parameter as it has a simple and direct physical meaning and is present with more stability in the entire range of relative humidity variation. Another favorable point about this model is that hysteresis effects are much less evident on transport coefficients associated to moisture content gradients than those associated to capillary pressure gradients. As a result, the disregard of hysteresis implies likely small errors when moisture content gradient is the driving potential.

Moreover, as a driving potential for moisture flow, some researchers have adopted gradients of relative humidity, vapor pressure, vapor density, capillary suction pressure, or porous matrix potential. Among these, Dos Santos and Mendes [170] used both vapor pressure and gas pressure gradients, Pedersen [157] used both capillary suction pressure and vapor pressure gradients, and Künzel [119] used the relative humidity gradient in the WUFI code. Their calculation methodologies appropriately consider this discontinuity phenomenon at the interface. Janssen discussed the choice of the driving potentials [102] and concluded that it depends on the level of moisture content accumulated in the material. Researchers have actually reformulated the Philip and De Vries model [162] by using different potentials rather than the potential that has the moisture content as the variable across the porous physical domain.

In the HAMSTAD project [91], a more complete model based on temperature and vapor pressure gradients was proposed to allow the calculation of HAM flows through building elements. The HAMSTAD energy conservation equation is

$$\frac{\partial}{\partial x}\left(\lambda \frac{\partial T}{\partial x}\right) - r_a \rho_a c_a \frac{\partial T}{\partial x} + \frac{\partial}{\partial x} L\left(\delta_v \frac{\partial \rho_v}{\partial x}\right) - r_a L \frac{\partial P_v}{\partial x} = c_m \rho_0 \frac{\partial T}{\partial t}, \qquad (2.14)$$

where the first, second, third, and fourth left-hand terms are respectively the net heat flux by conduction, advective flux due to the transport through the porous material, net heat flux due to phase change in the diffusive transport, and net heat flux due to phase change in the advective transport. The right-hand term corresponds to the transient term for the energy in the control volume. The parameter r_a is the volumetric air flow, ρ_a is the air density, and c_a is the specific heat at constant pressure.

The moisture balance was described by Hangentoft [91] as

$$\frac{\partial}{\partial x}\left(\delta_v \frac{\partial P_v}{\partial x}\right) - r_a L \frac{\partial \rho_v}{\partial x} - \frac{\partial}{\partial x}\left(\delta_l \frac{\partial P_c}{\partial x}\right) = \frac{\partial \omega}{\partial t}, \tag{2.15}$$

where the first, second, and third left-hand terms are respectively the net diffusive vapor flux represented by the Fick's law, net advective vapor flux, and net liquid water flux represented by Darcy's law. In addition, δ_l is the hydraulic conductivity, P_c is the capillary pressure, ω is the moisture content in terms of mass, and δ_v is the vapor permeability.

To calculate the air transport based on data of air permeability, Santos and Mendes [56] presented a three-governing equation model based on three driving potentials that consider the differential governing equations for energy, moisture, and dry-air balances, as follows:

$$c_m \rho_0 \frac{\partial T}{\partial t} = \nabla \cdot \left(\left(\lambda - \delta_l \frac{\partial P_c}{\partial T} c_l T\right) \nabla T - \left(\delta_l \frac{\partial P_c}{\partial P_v} c_l T + \delta_v c_a T - \delta_v (\mathscr{L} + c_v T)\right) \nabla P_v\right)$$
$$+ \nabla \cdot \left(\left(\rho_a \frac{k k_g}{v_g} c_a T + \rho_v \frac{k k_g}{v_g} L\right) \nabla P_g + K \rho_l c_l T g\right),$$

$$\frac{\partial \omega}{\partial \phi} \frac{\partial \phi}{\partial P_v} \frac{\partial P_v}{\partial t} + \frac{\partial \omega}{\partial \phi} \frac{\partial \phi}{\partial T} \frac{\partial T}{\partial t}$$
$$= \nabla \cdot \left[-\delta_l \frac{\partial P_c}{\partial T} \nabla T - \left(\delta_l \frac{\partial P_c}{\partial P_v} - \delta_v\right) \nabla P_v + \rho_v \frac{k k_g}{v_g} \nabla P_g + \delta_l \rho_l g\right],$$

$$\frac{\partial \rho_a}{\partial P_g} \frac{\partial P_g}{\partial t} + \frac{\partial \rho_a}{\partial P_v} \frac{\partial P_v}{\partial t} + \frac{\partial \rho_a}{\partial T} \frac{\partial T}{\partial t} = \nabla \cdot \left(-\delta_v \nabla P_v + \rho_a \frac{k k_g}{v_g} \nabla P_g\right),$$

where k is the absolute permeability (m^2), k_g is the gas relative permeability, v_g is the dynamic viscosity ($Pa.s$), P_g is the gas pressure ω, and P_g is the gas pressure (dry air pressure + vapor pressure, in Pa). One of the advantages of this three-driving potential model is to be more comprehensive and to consider the transient air balance. It is solved using the Multi–Tridiagonal–Matrix Algorithm (MTDMA) for a vector with three dependent variables.

This three governing equations model was also used for solving 2D problems associated to corners, roughs and foundations [55, 56, 171–174]. In all programs, porous building elements (walls, roofs, and furnishing) can be considered as composite components consisting of several layers of porous materials, each subdivided into several control volumes, for which the calculations are conducted according to the adopted numerical method. In each time step, temperature and another variable representing moisture are calculated in each control volume. Sorption isotherm curves are used to convert vapor pressure or relative humidity into moisture content, and normally an implicit calculation procedure is used to ensure numerical stability even for relatively large time steps.

To solve diffusion problems considering variable properties, dependent local sources, and moisture effects, simulation programs adopt numerical methods to more precisely compute the energy and mass balances for each zone and for the whole building.

2.3 Boundary Conditions

The boundaries of the physical domain are the interfaces between the surrounding fluid, air, and porous building element. If we consider the 1D assumption appropriate, we can write the following expression for the boundary conditions in the energy conservation equation:

$$
-\left[\lambda\left(T,\theta\right)\frac{\partial T}{\partial x}-L\,\mathbf{j}_v\right]_{x=0}=h_T\left(T_\infty-T_{x=0}\right)+Lh_{\rho v}\left(\rho_{v,\infty}\left(T,\theta\right)-\rho_{v,x=0}\left(T,\theta\right)\right),
$$
(2.16)

where the first right-hand term $h_T\left(T_\infty-T_{x=0}\right)$ represents the convective heat exchanged with the outside air and $Lh_{\rho v}\left(\rho_{v,\infty}-\rho_{v,x=0}\right)$ is associated to the phase change energy term. The short- and long-wave radiation terms can be included in Eq. (2.16) as follows:

$$
-\left[\lambda\left(T,\theta\right)\frac{\partial T}{\partial x}-L\,\mathbf{j}_v\right]_{x=0}=h_T\left(T_\infty-T_{x=0}\right)+Lh_{\rho v}\left(\rho_{v,\infty}\left(T,\theta\right)-\rho_{v,x=0}\left(T,\theta\right)\right)
$$
$$
+\alpha\mathbf{q}_r+\sum_{i=1}^{m}f\varepsilon\chi\left(T_i^4-T_{x=0}^4\right),
$$

where the first right-hand term represents the heat exchanged through convection, the second term is the phase change energy term L as the vapor–liquid latent heat, the third term is the absorbed short-wave radiation, and the fourth term is the long-wave radiation (m is the number of neighboring surfaces). The parameters f, ε, and χ are respectively the view factor, emissivity, and Stefan–Boltzmann constant ($5,67e^{-8}$ W/m$^2-$ K^4).

The boundary conditions related to long-wave radiative heat transfer can be linearised and written in a manner similar to the convective term, whereas the short-wave radiation is written as a source term in the energy-conservation boundary condition equation. For the mass conservation equation, we can write

$$-\left[\frac{\partial}{\partial x}\left(\left(D_{\theta_l}(T,\theta)+D_{\theta_v}(T,\theta)\right)\frac{\partial\theta}{\partial x}+\left(D_{T_l}(T,\theta)+D_{T_v}(T,\theta)\right)\frac{\partial T}{\partial x}\right)\right]_{x=0}$$

$$=\frac{h_{\rho v}}{\rho_l}\left(\rho_{v,\infty}(T,\theta)-\rho_{v,x=0}(T,\theta)\right).$$

$$(2.17)$$

The vapor concentration ρ_v depends on the temperature and relative humidity. At the external boundary surfaces, the relative humidity is calculated using the moisture content value in the sorption isotherm curve; this may neglect the hysteresis effect, assuming that the sorption and desorption curves are the same.

For the HAMSTAD model, we should add the air transport (r_a ρ_a c_a, and T_∞) and the rain ($\mathbf{j}_l$ c_l, and T_∞) terms, where $\mathbf{j}_l = -\delta_l\left(\frac{\partial P_c}{\partial x}\right)_{x=0}$ is the boundary condition associated to the water from the rain. This term could also be added in Eq. (2.16).

For the moisture flow, vapor transport is considered because of the difference between the partial vapor pressure in air and at the external and internal surfaces:

$$\mathbf{j} = h_{Pv}\left(p_{v,\infty} - p_{v,surf}\right),$$

where $\mathbf{j}$ is the density of moisture flow rate (kg/m^{2}s) and β_v is the surface coefficient of water vapor transfer (s/m) calculated from the Lewis' relation. In the three-driving potential equation [170], we can consider, for the dry-air conservation equation, a prescribed value for the gas pressure at the envelope surface, i.e.

$$P_{g,\infty} = P_{g,surf}.$$

A very uncertain parameter is the convective coefficient for both heat and mass transfer, which is very difficult to be evaluated and depends on geometry, air flow patterns, temperature difference, and surface rugosity. The main uncertainty in this boundary condition is related to the airflow, and is magnified because of the presence of obstructions such as furniture, neighboring buildings, and trees. In addition, the assumption that they are constant over a building's wall surface, reduces the accuracy of the whole-building model and underestimates the risk of mold growth.

Emmel et al. presented some external correlations for low-rise buildings [61] and Alamdari and Hammond presented correlations for internal surfaces [3] as follows:

$$h_T = \left\{\left[a\left(\frac{\Delta T}{d}\right)^p\right]^m + \left[b\left(\Delta T\right)^q\right]^m\right\}^{\frac{1}{m}},$$

where $a = 1.5$, $b = 1.23$, $p = 0.25$, $q = 0.33$, $m = 6.0$, and $d = 2.5$ (wall height).

The convective mass transfer coefficients $h_{\rho v}$ can be correlated to the convective heat transfer coefficients h_T by means of the Lewis relation as

$$h_{\rho v} = \frac{h_T}{\rho_a\, c_a\, Le^{2/3}},$$

where the following values are commonly adopted: $\rho_a = 1.166\,\text{kg/m}^3$, $c_a = 1.007\,\text{kJ/kg} - \text{K}$, and Le $= 1$.

The presence of paint layers has no storage effect but influences the absorptivity of the wall to solar radiation and the resistivity to the vapor flow that can provide a significant contribution to the energy and mass balance. This is because for some cases of painting, the convective mass coefficient $h_{\rho v}$ is considerably reduced along with the latent load. The paint is considered as an additional resistance to that created by the boundary layer.

Thus, the paint can be modeled as an additional resistance to boundary layer resistance to the water vapor transmission. The base normal SI unit for permeance p is kg/s/m^2/Pa, which is equivalent to $1.74784 \cdot 10^{10}$ US perms, so that the resistance added to the convective mass transfer coefficient associated to a vapor concentration difference is calculated as $\frac{RT}{Mp}$. For example, for latex, p can be considered as $190 \cdot 10^{-12}$ g/m/s/Pa for the external side and $980 \cdot 10^{-12}$ g/m^2/s/Pa for the internal side [26].

2.4 Discretization

For numerical simulation, the differential governing equations must be solved using a numerical method. In Chaps. 3 and 4, the finite-difference and finite-element methods are presented in detail; these methods are conventionally used to solve diffusion equations. In the second part of the book, Chaps. 5, 6, 7, and 8 present advanced methods and highlight their advantages.

This section briefly describes the discretization of the governing equations (Eqs. (2.13a), (2.13b), (2.16), and (2.17)) for a 1D problem, turning them into a system of algebraic equations, where each equation represents the balance for each point of the calculation domain. Figure 2.4 represents this domain and the system adopted for indexing the equations; the solid lines are located in the center of the control volumes and the dotted lines represent the boundaries between two control volumes.

The discretization procedure is that proposed by Patankar [155], using the technique of finite volume. Figure 2.3 shows the points on the elements of interest and the symbology. The capital letters represent the control volume (center point) and the small letters represent the boundaries between two volumes and follow the transport coefficients. These coefficients are calculated at the interfaces of volumes to represent the heat and mass transfer conditions along the distance separating two consecutive volume centers. Hence, one can deduce the following expression (harmonic mean) for calculating a certain coefficient $\mathscr{D}$ at an interface e:

$$\frac{\delta x_e}{\mathscr{D}_e} = \frac{\delta x_e^-}{\mathscr{D}_P} + \frac{\delta x_e^+}{\mathscr{D}_E},$$

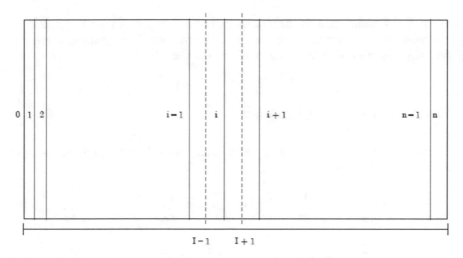

Fig. 2.3 Indexing of control volume elements in the calculation domain

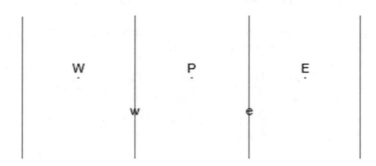

Fig. 2.4 Schematic grid for the domain discretization

where δx_e^- and δx_e^+ represent the distances between the face "e" and the points P and E, respectively (Fig. 2.4).

For a uniform grid, we can calculate a transport coefficient $\mathscr{D}$ at an interface by the following expression:

$$\mathscr{D}_e = \frac{2}{\dfrac{1}{D_P} + \dfrac{1}{D_E}} \ .$$

Or for the interface w as

$$\mathscr{D}_w = \frac{2}{\dfrac{1}{D_P} + \dfrac{1}{D_W}} \ .$$

In the finite-volume method, the equations in the time domain are discretized by integrating them over a time interval Δt. Moreover, the discretization in space is performed by integrating the equation over a volume element.

Discretized Governing Equations

By rewriting the mass and energy conservation equations for a $1D$ problem, we have the following differential equations:

$$\frac{\partial \theta}{\partial t} = \frac{\partial}{\partial x}\left(\mathrm{D}_T \frac{\partial T}{\partial x} + \mathrm{D}_\theta \frac{\partial \theta}{\partial x}\right) \tag{2.18}$$

and

$$\frac{\partial (\rho_0 c_m T)}{\partial t} = \frac{\partial}{\partial x}\left(\lambda \frac{\partial T}{\partial x}\right) + \mathscr{L}\rho_l \frac{\partial}{\partial x}\left(\mathrm{D}_{TV}\frac{\partial T}{\partial x} + \mathrm{D}_{\theta V}\frac{\partial \theta}{\partial x}\right) . \tag{2.19}$$

The discretization of Eqs. (2.18) and (2.19) gives the following algebraic equations for the internal points of the physical 1D domain:

$$\left(\frac{\Delta x}{\Delta t} + \frac{\mathrm{D}_{\theta e}}{\delta x_e} + \frac{\mathrm{D}_{\theta w}}{\delta x_w}\right)\theta_P = \frac{\mathrm{D}_{\theta e}}{\delta x_e}\theta_E + \frac{\mathrm{D}_{\theta w}}{\delta x_w}\theta_W + \frac{\Delta x}{\Delta t}\theta_P^{\mathrm{prev}}$$
$$+ \left[\frac{\mathrm{D}_{Te}}{\delta x_e}\left(T_E^{\mathrm{prev}} - T_P^{\mathrm{prev}}\right) - \frac{\mathrm{D}_{Tw}}{\delta x_w}\left(T_P^{\mathrm{prev}} - T_W^{\mathrm{prev}}\right)\right]$$

and

$$\left(\rho_0 c_m \frac{\Delta x}{\Delta t} + \frac{\lambda_e}{\delta x_e} + \frac{\mathscr{L}\mathrm{D}_{TVe}}{\delta x_e} + \frac{\lambda_w}{\delta x_w} + \frac{\mathscr{L}\mathrm{D}_{TVw}}{\delta x_w}\right) T_P =$$
$$\left(\frac{\lambda_e}{\delta x_e} + \frac{\mathscr{L}\mathrm{D}_{TVe}}{\delta x_e}\right) T_E + \left(\frac{\lambda_w}{\delta x_w} + \frac{\mathscr{L}\mathrm{D}_{TVw}}{\delta x_w}\right) T_W$$
$$+ \rho_0 c_m \frac{\Delta x}{\Delta t} T_P^{\mathrm{prev}} + \mathscr{L}\rho_l\left(\frac{\mathrm{D}_{\theta Ve}\left(\theta_E^{\mathrm{prev}} - \theta_P^{\mathrm{prev}}\right)}{\delta x_e} + \frac{\mathrm{D}_{\theta Vw}\left(\theta_W^{\mathrm{prev}} - \theta_W^{\mathrm{prev}}\right)}{\delta x_w}\right) ,$$

where the superscript "prev" indicates that the variable has a value for the previous time interval.

Similarly, for the boundary conditions with a half-control volume, the following expressions are obtained at the external surface ($x = 0$):

$$\left(\frac{\Delta x}{2\Delta t} + \frac{\mathrm{D}_{\theta e}}{\delta x_e}\right)\theta(0) = \frac{\mathrm{D}_{\theta e}}{\delta x_e}\theta(1) + \frac{\Delta x}{2\Delta t}\theta^{\mathrm{prev}}(0)$$
$$+ \mathrm{D}_{Te}\left(\frac{T^{\mathrm{prev}}(1) - T^{\mathrm{prev}}(0)}{\delta x_e}\right) + \frac{h_{\rho v}}{\rho_l}\left(\rho_{v,\mathrm{ext}} - \rho(0)\right)$$

and

$$
\left(\rho_0 c_m \frac{\Delta x}{2 \Delta t} + \frac{\lambda_e}{\delta x_e} + \mathscr{L} \rho_l \frac{\mathrm{D}_{TVe}}{\delta x_e} + h_T \right) T(0) =
$$

$$
\left(\frac{\lambda_e}{\delta x_e} + \mathscr{L} \rho_l \frac{\mathrm{D}_{TVe}}{\delta x_e} \right) T(1) + \left(\rho_0 c_m \frac{\Delta x}{2 \Delta t} \right) T^{\mathrm{prev}}(0)
$$

$$
+ \mathscr{L} \rho_l \mathrm{D}_{\theta Ve} \left[\frac{(\theta^{\mathrm{prev}}(1) - \theta^{\mathrm{prev}}(0))}{\delta x_e} \right] + h_T T_{\mathrm{ext}} + \alpha q_r \tag{2.20}
$$

$$
- \sum_{i=1}^{m} f \varepsilon \chi \left(T_i^4 - T_{x=0}^4 \right) + \mathscr{L} h_{\rho v} \left(\rho_{v,\mathrm{ext}} - \rho_v(0) \right).
$$

The long-wave radiation exchanged with outer neighboring surfaces was neglected in Eq. (2.20); however, it can be represented by the next-to-last right-hand term, which can include the deep-sky radiative heat transfer. The discretization for the internal surface is very similar, and this is not shown.

Traditionally, the discretised governing equations are placed in the form of Eq. (2.21) and are solved iteratively by using the Tridiagonal Matrix Algorithm (TDMA).

$$
A_P \Phi_P = A_E \Phi_E + A_W \Phi_W + D \tag{2.21}
$$

where

$$
A_P = A_E + A_W + A_0^0; \qquad D = A_P^0 \Phi_P^0 + F;
$$

Φ is a dependent variable (T or θ).

The source term D in Eq. (2.21) holds, beyond the transient term ($A_P^0 \Phi_P^0$), all source terms (F) that are nonlinear functions of the dependent variable.

The iterative-solution method allows the transport coefficients to be evaluated using the values of temperature and moisture content of the previous iteration (subscript $_{ant}$). Owing to the iteration procedure, convergence control criteria of the numerical method are needed. Normally, a relative difference between two consecutive iterations lower than 10^{-4} is a conservative criterion but depends on the material and time step. To illustrate the high dependability of moisture properties, Fig. 2.5 presents the variation of transport coefficients for lime mortar associated to a temperature gradient.

2.5 Stability Conditions

The use of the different models, boundary conditions, and the dependence especially on moisture content show a very strong coupling between the equations. A possible reason that has motivated researchers to numerically decouple the governing equations may be the difference between the time constants for the two diffusion phenomena (heat and moisture transfer).

Fig. 2.5 Moisture transport coefficient for lime mortar associated to a temperature gradient [158]

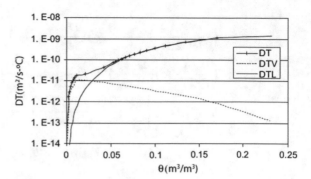

When the heat and moisture equations are decoupled for the numerical solution, high magnitude source terms are likely to appear. However, linear systems of equations are numerically solvable when the main diagonal terms are dominant. Thus, the source term must be dismembered so that its magnitude decrease and the main diagonal increases.

In addition, for models based on moisture content or capillary suction pressure gradients, a great amount of energy gain may occur due to phase change at a given step because of a given amount of vapor condensing on the surface, which may substantially increase the temperature and moisture content or the capillary suction pressure at the surface, thus numerically inverting the vapor flow. This unrealistic behavior occurs because the mass flow resistance toward the opposite side of the wall is higher and can occur continuously for greater time steps, thus preventing the solution from converging. Therefore, numerical stability with traditional algorithms, such as TDMA, can only be ensured for very small time steps.

To provide more stability and allow increase in the simulation time step, it is recommended that source terms are linearized and an algorithm, allowing the determination of all fields at the same iteration level, is used. For this combined heat and moisture transfer problem through a porous medium, Mendes et al. discretized the conservation equations by using the control-volume formulation method with a central difference scheme, and linearized the vapor concentration difference at the boundaries in terms of temperature and moisture content [140]. They then used a coupled algorithm (i.e., MTDMA) to obtain a more robust solution.

2.6 Linearization of Boundary Conditions or Source Terms

To decrease the magnitude of the source terms, the boundary conditions can be linearized, especially when the moisture model is based on gradients such as moisture content and capillary pressure.

For example, the discretization of the energy conservation equation at the boundaries can alternatively be given by

$$
\left(\rho_0 c_m \frac{\Delta x}{2\Delta t} + \frac{\lambda_e}{\delta x_e} + L\rho_l \frac{D_{TVe}}{\delta x_e} + h_T + Lh_{\rho v}C_l \right) T(S) =
$$
$$
\left(\frac{\lambda_e}{\delta x_e} + L\rho_l \frac{D_{TVe}}{\delta x_e} \right) T(n(S)) + \tag{2.22}
$$
$$
L\rho_l \frac{D_{\theta Ve}}{\delta x_e} (\theta^{\text{prev}}(n(S)) - \theta^{\text{prev}}(S)) + h_T T_\infty +
$$
$$
Lh_{\rho v} (C_l T_\infty + C_2 (\theta_\infty - \theta^{\text{prev}}(S)) + C_3) + \rho_0 c_m \frac{\Delta x}{2\Delta t} T^0(S),
$$

where $n(S)$ is the internal point neighbor to S and the coefficients $C1$, $C2$, and $C3$ are associated to the linearization of the vapor concentration difference in terms of temperature and moisture content [140]. Similar treatment can be applied, for instance, for capillary pressure difference or any other chosen driving potential. Equation (2.22) has a suitable form to be applied to TDMA. The same linearization can be applied to the mass conservation equation at the boundaries.

2.7 Numerical Algorithms

Solution algorithms normally used to solve the algebraic systems of governing equations can be direct or iterative. Linear systems of equations are usually solved by using well-known methods, such as Gaussian elimination and Gauss–Jordan methods (a variant of Gauss elimination scheme), and other direct methods such as Crout reduction and Cholesky methods [74]. These elimination methods use back substitution to obtain the solution; this is performed by solving the set of equations in reverse order as in the TDMA.

Nevertheless, linear systems may be very sensitive to roundoff errors created by direct-method procedures, and the solution might be inaccurate. Therefore, for heat transfer problems described by a system of equations with sparse coefficient matrices, iterative methods, such as Gauss–Seidel, Jacobi, and over-relaxation, may be more precise and rapid, requiring less computer resident memory.

A very convenient direct method for 1D problems is the aforementioned TDMA, which can be combined with the iterative Gauss–Seidel method to perform multidimensional calculation.

Nevertheless, it is well-known that linear systems of equations converge for any starting value when they are diagonally dominant, i.e., the diagonal entry of the coefficient matrix is larger than the sum of the other coefficients. This usually occurs in heat transfer problems; however, in certain situations, e.g., at the boundaries of combined heat and mass transfer problems in porous media, this condition may not be satisfied, thus making the solution very unstable.

Künzel used the alternating-direction implicit (ADI) method in WUFI [119], with successive iterations between heat and moisture conservation equations until reaching a desired accuracy, for modeling a 2D heat and moisture transfer problem through porous building components.

To use a robust algorithm to simultaneously solve the governing equations with strong coupling among the dependent variables, Mendes presented a TDMA-based procedure with a generalization for multivariable problems, called the MTDMA [137, 138].

This algorithm uses tensors instead of scalar numbers so that it rapidly reaches numerical convergence by simultaneously solving all equations at the same iteration level, thus being more numerically stable for much larger time steps than TDMA.

Coupled heat and mass transfer in porous media is very critical due to high divergence-related problems. This is described in the IEA Annex 24 (1996) and by Pedersen [157]. Wang and Hagentoft described a numerical method for calculating combined HAM transport in building envelope porous components [198]. Their algorithm combines an explicit scheme for heat and moisture transport and a relaxation scheme for air transport.

However, the maximum time step is calculated in terms of a stability criterion for heat and moisture transport, which could be greatly improved using the MTDMA, in a stable manner. In fact, both stability criteria and time step control devices are needed in this case to reduce numerical convergence problems.

In HAM models, heat, air, vapor, and liquid flow must be used simultaneously. Furthermore, physical quantities, such as mass transport coefficients, thermal conductivity, and specific heat, are variable and dependent on both temperature and moisture content. At the boundaries, heat transfer convection and vapor/liquid-phase change are considered. All these make the solution complex and lead to a highly-coupled system of nonlinear equations, considerably increasing computer run time for long-term simulations; in several cases, depending on the magnitude of boundary conditions, it might be impossible to reach numerical convergence, or the convergence is only achieved for small time steps. Therefore, solution algorithms and numerical methods must be explored and improved.

2.8 Multitridiagonal-Matrix Algorithm

Implicit schemes demand the use of an algorithm to solve tridiagonal systems of linear equations. One of the most popular algorithms used is the well-known Thomas Algorithm or TDMA. However, for strongly-coupled equations of heat transfer problems, a more robust algorithm may be necessary to provide numerical stability. Therefore, the MTDMA is used, especially for 1D problems.

For a physical problem represented by M dependent variables, the discretization of $M \times N$ differential equations generates the following system of algebraic equations:

$$\mathbf{A}_i \cdot \mathbf{x}_i = \mathbf{B}_i \cdot \mathbf{x}_{i+1} + \mathbf{C}_i \cdot \mathbf{x}_{i-1} + \mathbf{E}_i \,, \tag{2.23}$$

where $\mathbf{x}$ is a vector containing M−dependent variables ϕ_j (temperature, vapor pressure, moisture content, relative humidity, or any other physical quantity), and is represented as

$$\mathbf{x} = \begin{bmatrix} \phi_{1,j} \\ \phi_{2,j} \\ \vdots \\ \phi_{M,j} \end{bmatrix}.$$

Unlike in the traditional TDMA, coefficients $\mathbf{A}$, $\mathbf{B}$, and $\mathbf{C}$ in the MTDMA are $M \times N$ matrices, in which each line corresponds to one dependent variable. The elements that do not belong to the main diagonal are the coupled terms for each conservation equation. $\mathbf{E}$ is an M−element vector.

As MTDMA has the same essence of TDMA, it is necessary to replace Eq. (2.23) by relationships of the form:

$$\mathbf{x}_i = \mathbf{P}_i \cdot \mathbf{x}_{i+1} + \mathbf{q}_i, \tag{2.24}$$

where $\mathbf{P}_i$ is now a $M \times N$ matrix.

Similarly, vector $\mathbf{x}_i$ can be expressed in terms of $\mathbf{x}_{i+1}$,

$$\mathbf{x}_{i-1} = \mathbf{P}_{i-1} \cdot \mathbf{x}_i + \mathbf{q}_{i-1}.$$

Substitution of Eq. (2.24) into Eq. (2.20) gives

$$\mathbf{A}_i \cdot \mathbf{x}_i = \mathbf{B}_i \cdot \mathbf{x}_{i+1} + \mathbf{C}_i \cdot (\mathbf{P}_{i-1} \cdot \mathbf{x}_i + \mathbf{q}_{i-1}) + \mathbf{E}_i \tag{2.25}$$

or

$$(\mathbf{A}_i - \mathbf{C}_i \mathbf{P}_{i-1}) \cdot \mathbf{x}_i = \mathbf{B}_i \cdot \mathbf{x}_{i+1} + \mathbf{C}_i \cdot \mathbf{q}_{i-1} + \mathbf{E}_i. \tag{2.26}$$

By explicitly writing Eq. (2.25) for $\mathbf{x}_i$, we get

$$\mathbf{x}_i = \left[(\mathbf{A}_i - \mathbf{C}_i \mathbf{P}_{i-1})^{-1} \cdot \mathbf{B}_i \right] \cdot \mathbf{x}_{i+1} + (\mathbf{A}_i - \mathbf{C}_i \mathbf{P}_{i-1})^{-1} (\mathbf{C}_i \cdot \mathbf{q}_{i-1} + \mathbf{E}_i).$$

For consistency of the formulas, Eqs. (2.26) and (2.23) are then compared, generating the following recursive expressions:

$$\mathbf{P}_i = \left[(\mathbf{A}_i - \mathbf{C}_i \mathbf{P}_{i-1})^{-1} \cdot \mathbf{B}_i \right]$$

and

$$\mathbf{q}_i = (\mathbf{A}_i - \mathbf{C}_i \mathbf{P}_{i-1})^{-1} (\mathbf{C}_i \cdot \mathbf{q}_{i-1} + \mathbf{E}_i).$$

Once these matricial coefficients are calculated, the back substitution very mechanically provides all elements of vector $\mathbf{x}_i$.

The use of this algorithm makes the systems of equations more diagonally dominant, and its use is illustrated in the next section for the study of heat and mass transfer through porous media. The diagonal dominance is improved because the $\mathbf{A}_i$ coefficients are increased at the same time the $\mathbf{E}_i$ source terms are decreased.

The heat and mass transfer problem is highly nonlinear because of the high variation of moisture capacity, especially at high values of relative humidity, thus varying the transport coefficients two, three, or even more orders of magnitude, as shown in Fig. 2.5.

Therefore, with the use of MTDMA, iterations only occur because of

1. high moisture-content-dependent transport coefficients,
2. residuals from the linearization of boundary conditions, and
3. solutions of 2D or 3D problems.

2.9 Mathematical Model for a Room Air Domain

A lumped formulation for both temperature and water vapor mass can be adopted for the internal air domain of a building room coupled to the diffusion model. Equation (2.27) describes the energy conservation applied to the control volume involving the room air, which is submitted to loads of conduction, air infiltration and ventilation, and heat again.

$$\dot{E}_t + \dot{E}_g = \rho_a c_a V_a \frac{dT}{dt}, \tag{2.27}$$

where $\dot{E}_t$ is the energy flow that passes through the room (W), $\dot{E}_g$ is the internal energy generation rate (W), ρ_a is the air density (kg/m^3), c_a is the specific heat of air (J/kg $-$ K), V_a is the room volume (m^3), and T is the room air temperature ($^\circ$C).

The term $\dot{e}_t$ includes loads associated to the building envelope and latent conduction heat transfer, fenestration (conduction and solar radiation), and openings (ventilation and infiltration).

In terms of water vapor mass balance, different contributions can be considered: ventilation, infiltration, internal generation, people's breath, and floor surface. The lumped formulation becomes

$$\rho_a V_a \frac{dW}{dt} =$$
$$(\dot{m}_{\text{inf}} + \dot{m}_{\text{vent}}) (W_{\text{ext}} - W) + \dot{m}_b + \dot{m}_{\text{ger}} + \sum_{j=1}^{n} h_{\rho v} A_{j,f} \left[W_{v,f} - W \right], \tag{2.28}$$

where $\dot{m}_{\text{inf}}$ is the mass flow by infiltration (kg/s), $\dot{m}_{\text{vent}}$ is the mass flow by ventilation (kg/s), W_{ext} is the external humidity ratio (kgwater/kgdryair), W is the internal humidity ratio (kgwater/kgdryair), $\dot{m}_b$ is the flow of water vapor from the breaths of occupants (kg/s), $\dot{m}_{\text{ger}}$ is the internal water-vapor generation rate (kg/s), $h_{\rho v}$ is the mass transfer coefficient (kg/m^{2}s), $A_{j,f}$ is the area of the jth control volumes of the floor surface (m^2), and $W_{v,f}$ is the humidity ratio of each control volume (kg of water/kg dry air).

Santos and Mendes presented and discussed different numerical methods used to integrate the differential governing equations in the air domain (2.27) and (2.28), showing the results in terms of accuracy and computer run time [54]. In their analysis, they showed that the use of explicit methods, such as Euler and modified Euler, requires imperatively very small time steps, making simulations extremely time consuming. A third method used was obtained through a MATLAB program, and provides analytical expressions to be solved in a real simultaneous manner. However, these expressions are time consuming due to their large size, even though they require fewer iterations because of their numerical robustness. Santos and Mendes proposed the use of a semi-analytical method to solve the differential governing equations for the room air, as it combines robustness and rapidity [54]. This last method analytically solves each equation (mass and energy balances) through a numerical coupling among them. As a result, for the proposed problem, the room air balance equations can be written as

$$A + BT = \rho c V \frac{\mathrm{d}T}{\mathrm{d}t}$$

and

$$C + DW = \rho V \frac{\mathrm{d}W}{\mathrm{d}t},$$

which gives

$$T^{n+1} = \frac{e^{\frac{B}{\rho_a c_a V_a} \Delta t} (A + BT^n) - A}{B}$$

and

$$W^{n+1} = \frac{e^{\frac{D}{\rho_a V_a} \Delta t} (C + DW^n) - C}{D},$$

where T^n and W^n are the temperature and humidity ratio respectively, calculated at the previous time step, and

$$A = \sum_{i=1}^{m} h_T A_i T_i + \dot{E}_{gs} + \dot{E}_{gl}(W_{\text{prev}}),$$

$$B = -\sum_{i=1}^{m} h_T A_i,$$

$$C = \dot{m}_{\text{inf}} W_{\text{ext}} + \dot{m}_{\text{resp}}(T_{\text{prev}}) + \dot{m}_{\text{ger}} + \sum_{j=1}^{n} h_D A_{j,f} W_{v,f},$$

$$D = -\left(\dot{m}_{\text{inf}} + \dot{m}_{\text{vent}} + \sum_{j=1}^{n} h_{\rho v} A_{j,f} \right).$$

where $\dot{E}_{gs}$ is the sensible energy (infiltration + generation; W), $\dot{E}_{gl}$ is the latent energy (infiltration + generation; W), and T_i is the temperature of each surface of the building envelope (°C). The solution method for ordinary differential equations (ODE) is extensively treated in Chap. 3.

2.10 Hygrothermal Models Used in Some Available Simulation Tools

Besides the models detailed earlier, the reader might be interested in more information about other computer simulation models widely used for the prediction of heat and moisture transfer in building physics.

MATCH Model

MATCH is a 1D, transient model used for calculating combined heat and moisture transport in composite constructions. In this, the material parameters may vary with moisture content and temperature. Possible transfer of latent heat may be considered, along with hysteresis in the moisture retention curves. The energy balance is written in terms of specific enthalpy as

$$\rho_0 \frac{\partial h}{\partial t} = -\frac{\partial q}{\partial x} - L\frac{\partial j_v}{\partial x},$$

which is similar to Eq. (2.12).

For moisture balance, liquid and vapor transport are considered:

$$\rho_0 \frac{\partial \omega}{\partial t} = -\frac{\partial j_v}{\partial x} - \frac{\partial j_l}{\partial x},$$

where ω represents the sum of moisture in all the three phases: ice, liquid, and vapor (the ice phase is considered immobile).

As all other models, the heat flux is calculated using Fourier's Law and by allowing the thermal conductivity to vary with moisture content and temperature. The vapor flow is calculated using the Fick's law as

$$j_v = -\delta_v(\phi)\frac{\partial P_v}{\partial x}.$$

The liquid flow is calculated using Darcy's law as

$$j_l = \delta_l(\phi)\frac{\partial P_c}{\partial x}.$$

According to Pedersen [157], P_c is the capillary pressure, which is the opposite of the hydraulic pressure (usually negative). These pressures are assumed to be only dependent on moisture content, i.e., there is no influence of gravity or other external forces including osmotic pressures. Owing to extreme gradients encountered both in the suction pressure and hydraulic conductivity for even small moisture content gradients, an alternative formulation of Darcy's law was proposed by Pedersen [157] by using the logarithm of the suction pressure as the driving potential:

$$j_l = \delta_l P_c \frac{\partial \ln P_c}{\partial x}.$$

For the governing balance equations, one type of variable is used as the driving potential for a transport, while another variable is monitored in the corresponding balance equation. For instance, for heat transport: temperature and enthalpy; for moisture transport: vapor pressure and moisture content, or suction pressure and moisture content. Equations of state are used in each case to bind the driving potential to the variable in the balance equation.

The differential equations are numerically solved using the finite-difference method applied to a maximum of 30 layers in up to 12 materials. The solution is semi-implicit, and more details can be found in [157].

DELPHIN Model

Grunewald et al. [85, 87] presented a mathematical model comprising differential equations to predict heat and moisture transport, which led to the creation of the first version of the DELPHIN program, which is continuously developed at TUD, Germany. The model assumed conserved quantities defined as densities in relation to a representative elementary volume (REV). In the hygrothermal model, the change of the moisture mass comprises liquid water m_l and water vapor m_v. The corresponding moisture mass density is denoted as ρ_{l+v}.

Furthermore, in the hygrothermal model, a density can be formulated with respect to a specific phase volume, regarding either the liquid phase l or gas phase v. The density of the water vapor component with respect to the gas phase is denoted by the symbol ρ_{vg} and often referred to as vapor concentration. The moisture mass density can change according to mass fluxes or via imposed sources and sinks σ_{imp}. By considering the mass flux due to liquid capillary transport $\mathbf{j}_v$, water vapor diffusion $\mathbf{j}_{v,diff}$ and vapor convection with the bulk gas phase $\mathbf{j}_{v,\text{conv}}$, and the prescribed liquid water and water vapor sources σ_l and σ_v respectively, the mass balance equation is written as

$$\frac{\partial (\rho_{l+v})}{\partial t} = -\nabla \cdot \left(\mathbf{j}_l + \mathbf{j}_{v,\text{diff}} + \mathbf{j}_{v,\text{conv}} \right) + \sigma_{l,\text{imp}} + \sigma_{v,\text{imp}}^v.$$

The energy balance equation is formulated for the energy density u in [87]. The rate of change of the energy density depends on heat conduction flux $\mathbf{q}$, enthalpies h associated with each mass transport term, and enthalpy production associated with mass sources/sinks. The differential energy balance equation combines all these terms, with h_w, h_v, and h_a as specific enthalpies associated with liquid water, water vapor, and dry air, respectively:

$$\frac{\partial u}{\partial t} = -\nabla \cdot \left(\mathbf{q} + h_l \mathbf{j}_l + h_v \left(\mathbf{j}_{v,\text{diff}} + \mathbf{j}_{v,\text{conv}}\right) + h_a \mathbf{j}_{a,\text{conv}}\right) + h_l \sigma^l + h_v \sigma^v + \dot{u}_{\text{imp}}.$$

The two balance equations are complemented by constitutive equations that relate mass and energy densities with state variables and potentials, and flux, source, and sink terms with the corresponding transport functions and driving potentials. Further information and details can be found in [10].

WUFI Model

This model is based on the hygrothermal envelope calculation model by Künzel and Kiessl [119] that is also used in WUFI Plus and EnergyPlus. The following equations present the relationship for the temperature and relative humidity fraction:

$$\rho_0(c_0 + c_l \omega_l + c_v \omega_v) \frac{\partial T}{\partial t} = \frac{\partial}{\partial x}\left(\lambda \frac{\partial T}{\partial x}\right) + L \frac{\partial}{\partial x}\left(\delta_v \frac{\partial \phi}{\partial x} P_s\right),$$

$$\frac{\partial \omega}{\partial \phi} \frac{\partial \phi}{\partial t} = \frac{\partial}{\partial x}\left(D_l \frac{\partial \omega}{\partial \phi} \frac{\partial \phi}{\partial x}\right) + \frac{\partial}{\partial x}\left(\delta_v \frac{\partial \phi}{\partial x} P_s\right),$$

where δ_v is the vapor permeability in materials defined by the vapor diffusion resistance factor μ explained before, ϕ is the relative humidity, P_s is the saturation vapor pressure, $\frac{\partial \omega}{\partial \phi}$ is the moisture storage capacity, and D_l is the liquid transport coefficient.

The model spatial discretization algorithm, implemented in EnergyPlus, determines the number of nodes required with no more than 12 cells per material, with thinner widths near the boundaries where most changes are expected. To determine the temperature and relative humidity profiles of the next time step, the solver computes by iterating the heat and mass balance equations until convergence is reached. For this model, the time step must be as short as possible; in particular, if the building has frequent and large changes in internal and external temperatures or if driving rain is considered. The required properties for the latent heat exchanges are as follows.

- Porosity [m^3/m^3], which determines the maximum volumetric water content;
- Moisture sorption isotherm curve in [kg/m^3]: function of relative humidity fraction;

- Thermal conductivity curve function of moisture content [kg/m^3];
- Vapor diffusion resistance factor μ [-]: function of relative humidity fraction;
- Liquid transport coefficient D_w [m^2/s]: function of moisture content [kg/m^3] composed by suction and redistribution terms. These are selected according to the rain indicator of the weather file, and may be suction values when the exterior surface is wet due to rain or redistribution values when the surface is no longer wet;
- Initial moisture content [kg/kg].

WUFI is a family of software products, under continuous development, used for performing coupled heat and moisture calculations under local climate conditions and for materials, multilayer components, and whole buildings. Further information and details can be found in [67].

HAM-Tools

HAM Tool is an integrated simulation tool developed for heat, air, and moisture transfer analysis in building physics. The tools are designed as graphical block diagrams by using the graphical programming language Simulink as a modular structure of standard building elements. The software is open source, free of charges, and includes the following HAM model [9, 109]:

$$
\frac{1}{\rho_0} \frac{\partial \omega}{\partial \theta} = -\frac{\partial}{\partial x} \left(-\delta_l \frac{\partial P_c}{\partial x} - \delta_v \frac{\partial P_v}{\partial x} + \mathbf{j}_a \omega \right)
$$

$$
\rho_0 c_m \frac{\partial T}{\partial t} = -\frac{\partial}{\partial x} \left(-\lambda \frac{\partial T}{\partial x} + \mathbf{j}_a c_a T + L \mathbf{j}_v \right).
$$

This model is implemented using Simulink. In addition, new models implemented by users can be added to the software.

CAR-HAM

One of the main characteristics in this model is the inclusion of the radiative equation for obtaining more precision in the heat transfer through insulation materials that can be considered as a scattering medium. The governing balance equations are as follows:

$$
\frac{\partial \omega}{\partial \phi} \frac{\partial \phi}{\partial P_v} \frac{\partial P_v}{\partial t} + \frac{\partial \omega}{\partial \phi} \frac{\partial \phi}{\partial P_s} \frac{\partial P_s}{\partial T} \frac{\partial T}{\partial t} =
$$

$$
- \nabla \cdot \left[-\delta_v \nabla P_v + r_a \frac{P_v M}{\rho_l R T} + \delta_l \left(\frac{\partial P_c}{\partial P_v} \nabla P_v + \frac{\partial P_c}{\partial T} \nabla T - \rho g \right) \right]
$$

$$\frac{\partial \left[(\rho_0 c_0 + \omega_l c_l + \omega_v c_v)\, T \right]}{\partial t} = -\nabla \cdot \left[\left(-\lambda + \delta_l \frac{\partial P_c}{\partial T} h_l \right) \nabla T \right] - \nabla \cdot \left[\mathbf{j}_a \left(\frac{M P_v}{\rho_l R T} h_v + \rho_0 c_0 T \right) \right]$$

$$- \nabla \cdot \left[\left(\delta_l \frac{\partial P_c}{\partial P_v} h_l - \delta_v h_v \right) \nabla P_v \right] - \nabla \cdot \left[-\delta_l\, \rho g h_l \right] - \nabla \cdot \mathbf{q}_r$$

where δ_l is the liquid permeability, $P_c = \rho_l g \Psi$ is the capillary pressure, δ_v is the vapor permeability, and r_a is the air flow rate. The energy equation is computed according to the current HAMSTAD equation [91], except for the internal radiative flux. The divergence of the radiative heat flux $\nabla \cdot \mathbf{q}_r$ appears as a source term and must be solved sequentially by using the Radiative Transfer Equation (RTE), which is an integro-differential equation whose solution requires special treatment by being numerically solved using the discrete ordinate method:

$$\mu_i \frac{d I_i}{d\tau} = (1 - \beta) I_b - I_i + \frac{\beta}{4\pi} \sum_{j=1}^{n} w_j I_j \varphi_j \, .$$

In short,

$$\mu_i \frac{d I_i}{d\tau} + I_i = \sigma_{r,i} \, ,$$

with,

$$\sigma_{r,i} = (1 - \beta) I_b + \frac{\beta}{4\pi} \sum_{j=1}^{n} w_j I_j \varphi_j \, ,$$

where σ_r is the radiative source term, μ is the direction cosine, I is the intensity, τ is the optical thickness, β is the albedo, I_b is the black body intensity, w is the quadrature weight, and φ is the phase function.

The divergence of the radiative heat flux can be calculated using the field of intensities as:

$$\nabla \cdot \mathbf{q}_r = \alpha \left(4\pi I_b - \sum_{j=1}^{n} I_j \omega_j \right) \, ,$$

where α is the adsorption coefficient.

The model is solved using the MTDMA algorithm at each iteration level, and is included in the Domus simulation software. The radiative heat flow in fibrous materials, such as glass wool, on hot days might be more significant than conductive flux, as shown in [73]. Moreover, this model and derived models have been implemented using Fluent UDF (User Defined Functions), allowing 3D diffusion modeling in complex geometries to be coupled with a differential formulation for the air flow (CFD). The CFD-coupled model has been called CFD-HAM, which allows 3D simulation of diffusive heat and moisture transport in complex geometries combined with 3D airflow simulation as shown in Fig. 2.6. A more detailed description of this model can be found in [136]. Further information is presented in [163].

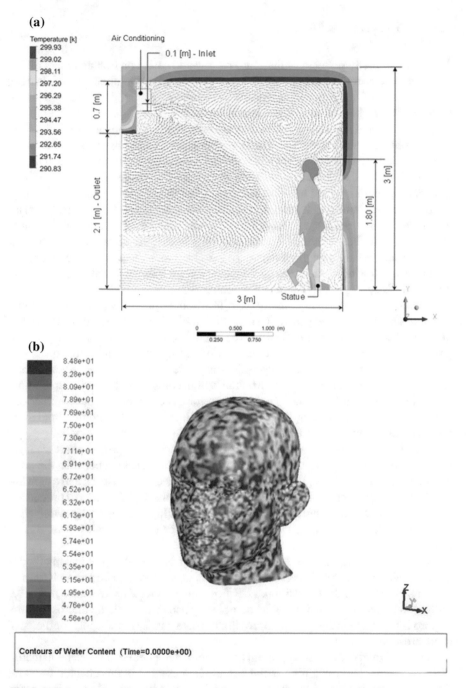

Fig. 2.6 Example of results obtained with CFD–HAM simulation from [136]

2.11   Final Remarks

This chapter briefly presented a literature review and basics on heat and mass diffusion phenomena applied to building physics. Despite great evolution over the last 30 years in terms of development of algorithms, mathematical models, and simulation tools, diffusion phenomena analysis is still limited.

Convective and radiative heat transfers are considered as boundary conditions but rarely within the porous structure. The same limitation is also observed regarding air flow. As the dimensions of height and width of a building are normally much larger than its thickness, with focus mainly on the highest moisture and vapor pressure gradients appearing within building envelopes, building codes commonly simplify the phenomenon, studying it unidimensionally. As such, the convective heat flux, as a boundary condition, at the surfaces is considered uniform over the entire heat and mass transfer surface. This hypothesis completely underestimates mold growth risks at the corners, where they often occur. To provide more accuracy to the boundary conditions, diffusion models should consider the multidimensional nature of heat and mass transport through building elements.

Additionally, thermal asymmetry is often noticed at the surface of building envelopes but is normally ignored in simulation tools. The multidimensionally combined heat and moisture transfer in large domains, such as grounds, may be of paramount importance for the building energy and hygrothermal evaluation; however, it is frequently neglected because of complexities, such as nonlinearities, highly variable transport coefficients, and particularly high computer run time for simulating whole buildings and large 3D domains. Therefore, the use of advanced numerical methods might be promising to advance research and use of building simulation codes, by avoiding limitations concerned especially to the simulation cost.

In addition, models must be improved to consider more proper phase-change mechanisms and the details of moisture transport in the transition between the pendular and funicular regions. Research is certainly needed in this aspect, and a considerable difference in terms of results might occur due to the high variation of properties with extremely small changes on relative humidity at high values of this parameter. In addition, there is a considerable contact resistance between wall layers; this is rarely considered.

As shown in Fig. 2.1, we believe research on advanced numerical methods, together with the progress in computer hardware, may allow applied research toward the urban physics scale. In addition, HAM models in micro/macro scales might be used for a better understanding of the complex transport through the microstructure of porous building elements, thus improving the data quality of transport coefficients of macroscopic models.

Chapters 3 and 4 respectively present the finite-difference and finite-element methods to solve the governing differential equations presented in this chapter. In addition, corresponding exercises are presented in these two chapters. Chapter 5 presents improved schemes that can be successfully applied to building simulation tools. Next, Chap. 6 provides the basics on ROMs and introduces some exercises. Although the use of these advanced methods are very recent in this field, the first results on the reduction of computer run time and low-cost storage have been shown as very promising for building simulation. The two other advanced methods presented in Chaps. 7 and 8 are also promising, especially for the modeling of large 3D domains, which can be a trend for district energy simulation or urban physics. Chapter 9 presents a dedicated section on exercises about heat and moisture diffusion problems in building physics. Chapter 10 provides final remarks and observation about the whole book.

In general, these observations attempt to motivate Ph.D. students to conduct research on building physics to improve the modeling and solution of diffusion phenomena occurring on a regular basis in building structures.

Chapter 3
Finite-Difference Method

This chapter is entirely devoted to numerical methods, keeping in mind that the main application is on diffusion processes in building physics. However, the presentation is oriented to the practical construction of numerical schemes with an overview of their elementary numerical properties.

As the subject has been widely studied, the literature could be quasi-infinite. For ordinary differential equations (ODEs), the reader may consult, for example, [30]. Classical references, considered as a type of *Bible* on the subject, include [92–94], which consider more sophisticated techniques. The presentation of Runge–Kutta (RK) methods can be found, for example, in [8, 29, 53, 193]. For partial differential equations (PDE), numerous books can be cited, including [185, 186]. For examples and applications on diffusion processes, the reader may consult [122, 169]. Furthermore, a presentation of numerical properties can be found, for example, in [28, 89], and a more general presentation on scientific computation can be found in [95]. An historical survey has been conducted in [188]. Finally, a very pedagogical and detailed reference on the finite-difference method for both ODE and PDE can be found in [126]. The aforementioned references can be completed according to preference and viewpoints.

Diffusion processes are governed by physical phenomena with characteristic scales, depending on the configuration and medium through which the transfer occurs. For example, thermal diffusion through building material is associated to a typical diffusion coefficient of order at $10^{-7}\,\mathrm{m^2/s}$, whereas the transport coefficient of moisture diffusion through soils can vary from 10^{-12} to $10^{-4}\,\mathrm{m^2/s}$. Moreover, the order of magnitude of these coefficients may drastically change as the diffusion process progresses, depending on the evolution of the temperature and moisture fields. Thus, to treat the largest number of configurations the models must be set into a non-dimensional formulation. This implies that the following typical scales are determined: the diffusion coefficient $\alpha^\star$, space scale $L^\star$, and thus typical time scale $\tau = (L^\star)^2/\alpha^\star$, and the typical order of magnitude of the dependent variable

© Springer Nature Switzerland AG 2019

N. Mendes et al., *Numerical Methods for Diffusion Phenomena in Building Physics*,
https://doi.org/10.1007/978-3-030-31574-0_3

u^*, which often indicates temperature. Therefore, for numerical considerations, the time step Δt and the space steps Δx, Δy, and Δz must remain lower than 1 (but strictly positive), whereas the dependent variable u lies in the order range of unity.

3.1 Numerical Methods for Time Evolution: ODE

This section presents some basic tools for numerical simulation related to time evolution problems. In addition, it is assumed that no space distinction is considered by the mathematical model. In technical words, we are dealing with Initial Value Problems (IVP) for finite dimensional systems, described by ODEs. For numerical purposes, one of the main issues is to estimate what can be the *optimal* time step (or at least to understand its selection) for a simulation over a fixed period. The solution is obtained based on the following:

1. The global error that is tolerated at the end of the simulation,
2. The appropriate behavior of the numerical solution in practice.

These two features are directly quantifiable (as theoretical estimations) because of some numerical properties: the accuracy of the method (related to its local truncation error) and its *absolute* stability.

The next section introduces some basics in numerical analysis with the help of a simple but fundamental example. Next, an overview is presented, concerning some results associated to some popular numerical methods in the general ODE case by using notions that have been introduced.

The literature references could be considered as quasi-infinite from a very theoretical viewpoint to practical implementation processes. Instead of providing several references, let us cite only the following:

- Hairer, E., Nørsett, S.P. & Wanner G., *Solving Ordinary Differential Equations I. Nonstiff Problems.* Springer [93]
  This is a type of Bible on the subject. It is at least one of the main reference books every researcher in the field must have. Notice that Part II discusses the numerical approaches to stiff problems relevant to building physics problems.
- LeVeque, R.J., *Finite Difference Methods for Ordinary and Partial Differential Equations.* SIAM [126]
  This is a complete and very pedagogical manuscript, with numerous examples and interpretations that make reading and learning easier.

3.1.1 An Introductory Example

In [54], the model of time evolution for the average room temperature T, as shown in Chap. 2, Sect. 2.9, is given by the following scalar ODE:

$$\rho_a c_a V_a \frac{dT}{dt} = -B\,T + A, \tag{3.1}$$

where the constants B and A include the surrounding effects that influence the temperature evolution along the time.

In the following text, we use this equation for applying the simplest and natural methods of discretization for ODE, introducing the fundamental concepts in numerical analysis.

3.1.1.1 A Fundamental Example

Consider the following problem

$$u'(t) = -\lambda u(t), \tag{3.2}$$

$$u(t^0) = u^0, \tag{3.3}$$

where $\lambda > 0$ and u_0 can be any fixed real numbers. The expression of the exact solution for this example is clearly $u_e(t) = u^0 e^{-\lambda t}$. For the present introductory example, we use the exact solution to highlight some numerical behavior associated with the schemes we constructed. The generalization is done in the next sections.

Exercise 2 Find the change of variable such that Eq. (3.1) is recast in the generic form of Eq. (3.2).

Answer: $\lambda = \dfrac{B}{\rho_a c_a V_a}$ and $u = T - \dfrac{A}{B}$.

3.1.1.2 Euler Schemes

The main goal for computing a numerical solution for a given system, e.g. Eq. 3.2, is to construct a recurrent sequence such that, by using at first the initial condition, we can step-by-step construct the next numerical approximation of the exact solution.

This recurrent sequence is called a *numerical scheme*; it is clearly not unique, and there are some fundamental properties that characterize each numerical scheme:

- Its *accuracy*: how far is the numerical approximation from the exact solution.
- Its *stability*: will the numerical scheme produce some numerical wriggles, oscillate indefinitely, or even explode to infinity at a finite time (even though it is not expected to)?
- Its *convergence*: generally speaking, does the numerical solution approach the exact solution as the time step decreases?

The first fundamental step for constructing such a numerical scheme is to fix a discretization for the independent variable time t and for the dependent variable u. Thus, assume a discretization of the time line $t^0, t^1, \ldots, t^n, t^{n+1}, \ldots$, such that

Fig. 3.1 Numerical
approximation
$u^0, u^1, \ldots, u^n$ of function
$u(t)$ at times $t^0, t^1, \ldots, t^n$

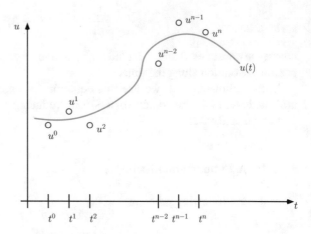

$t^{n+1} - t^n = \Delta t_n$. Next, we may assume that the time points are regularly spaced, so
that Δt is defined as a constant. Let u^n be the numerical approximation of $u(t^n)$, for
$n \geq 0$ (Fig. 3.1). The numerical scheme corresponds to a recurrent sequence such
that, given the values of $u^0, u^1, \ldots, u^{n-1}$, one can compute the next value u^{n+1}.
Owing to the well-posed problem, the initial value u^0 is always known, to allow us
to always initiate the recurrence.

For Eq. (3.2), the right-hand side (RHS) evaluated at a given time t^n is naturally
approximated by

$$-\lambda u(t^n) \simeq -\lambda u^n.$$

The approximation of the differential operator at the left-hand side (LHS) constitutes
the basis of the approach. The linear approximation of the first-order derivative
evaluated at time t^n can be

$$u'(t^n) \simeq \frac{u^{n+1} - u^n}{\Delta t}, \tag{3.4a}$$

$$u'(t^n) \simeq \frac{u^n - u^{n-1}}{\Delta t}. \tag{3.4b}$$

Equation (3.4a) corresponds to an approximation of the derivative by using a forward
discrete point u^{n+1}, whereas Eq. (3.4b) uses a backward discrete point u^{n-1}, as shown
in Fig. 3.2.

Thus, these two methods for discretizing the differential operator yield the following two numerical schemes for Eq. (3.2):

$$\frac{u^{n+1} - u^n}{\Delta t} = -\lambda u^n \qquad \text{Forward Euler (FE) scheme}$$

$$\frac{u^n - u^{n-1}}{\Delta t} = -\lambda u^n \qquad \text{Backward Euler (BE) scheme}$$

Fig. 3.2 Approximation of
$u'(t^n)$ using a forward
approximation *in brown* and
a backward approximation *in
blue*

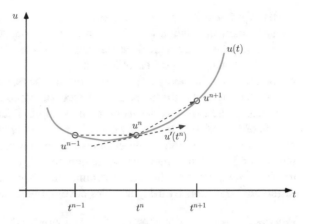

A slight shift of the index for the BE scheme highlights the differences between these
two schemes:

$$\frac{u^{n+1} - u^n}{\Delta t} = -\lambda u^n \quad \text{(FE scheme)}, \quad \text{and} \quad \frac{u^{n+1} - u^n}{\Delta t} = -\lambda u^{n+1} \quad \text{(BE scheme)}.$$

Let us examine the numerical behavior of these two recurrent sequences that can
be recast in the following form:

$$u^{n+1} = (1 - \lambda \Delta t)u^n \qquad \text{Forward Euler scheme},$$

$$u^{n+1} = \frac{u^n}{1 + \lambda \Delta t} \qquad \text{Backward Euler scheme}.$$

Recall that given u^0, we can compute u^1, which is the numerical approximation of
u at time $t^1 = t^0 + \Delta t$.

Accuracy

For both cases (FE and BE schemes), the approximation of the time derivative is
linear:

$$u'(t^n) = \frac{u^{n+1} - u^n}{\Delta t} + \mathcal{O}(\Delta t), \qquad u'(t^n) = \frac{u^n - u^{n-1}}{\Delta t} + \mathcal{O}(\Delta t).$$

The neglected terms are of the order of at least $\mathcal{O}(\Delta t)$, so that the highest value term
is proportional to Δt because the time step must be small (at least <1). Thus, for
both cases, the numerical scheme is said to be accurate up to order 1, or accurate at
$\mathcal{O}(\Delta t)$.

The signification is rather clear: the error committed at each iteration of the recurrence process is due to the truncated terms in the differential operator approximation.

With the progress of time, these errors are accumulated. It may then produce an inaccurate numerical solution for a long-time simulation. A method for reducing these time-accumulated errors involves reducing the value of the time step Δt, thus taking advantage of the truncation error being proportional to a power of Δt. However, this method may be extremely time consuming in case of long simulation periods.

Another method is to choose a more accurate approximation for the time derivative, that is, the neglected terms in the new approximation must be at least up the second order: $\mathcal{O}(\Delta t^2)$. Indeed, by using a first-order approximation for u' and a time step $\Delta t = 10^{-1}$, the truncation errors would be proportional to 10^{-1}. In contrast, when using a second-order approximation for u', the errors become proportional to 10^{-2}. Next, we show how to determine more such accurate approximations for differential operators by using a Taylor development method.

Remark 1 From the aforementioned, the accuracy is related to a local error of truncation of the differential operator.

Stability

This is another property expected by a numerical scheme to produce solutions without spurious behaviors, or even those that do not unexpectedly blow up. Viewed as recurrence sequences, both numerical schemes can be recast in the general geometric sequence as

$$u^{k+1} = q u^k, \qquad k \in \mathbb{N}.$$

Recall that when $|q| > 1$, the series $(u^k)_{k \in \mathbb{N}}$ tends to infinity in absolute value, as soon as $u_0 \neq 0$. This corresponds to an unstable behavior of the computed numerical solution. Thus, it is necessary to ensure the stability condition for the recurrence sequence

$$|q| \leqslant 1.$$

- For the FE scheme associated to Eq. (3.2), the stability condition reads

$$|1 - \lambda \Delta t| \leqslant 1,$$

  that is,

$$0 \leqslant \lambda \Delta t \leqslant 2.$$

- For the BE scheme, the stability condition becomes

$$\left| \frac{1}{1 + \lambda \Delta t} \right| \leqslant 1.$$

For any strictly positive value of time step Δt, as soon as $\lambda \geqslant 0$, the following inequality is satisfied: the BE scheme for Eq. (3.2) is said to be stable.

Exercise 3 Consider the scheme for Eq. (3.2), depending on the real parameter $\theta \in [0; 1]$:

$$\frac{u^{n+1} - u^n}{\Delta t} = -\lambda \left(\theta u^n + (1 - \theta) u^{n+1} \right) \qquad \theta\text{-scheme}$$

Then, what is the stability condition in the function of parameter θ?

> *Answer:*
> $$\left| \frac{1 - \theta \lambda \Delta t}{1 + (1 - \theta) \lambda \Delta t} \right| \leqslant 1 .$$

Remark 2 Here, we introduced a particular notion of stability, which is often called *A-stability*. A weaker notion of stability is the *zero-stability* that is rather a mathematical concept, which is not very useful in practice.

Consistency

For producing numerical schemes, we approximate the differential operator such that the truncation error tends to be 0 as the time step also tends to be 0. This can be formulated as the numerical scheme tends toward the differential equation when the discretization of the time line tends toward its continuous limit. This corresponds to the notion of consistency of the scheme with the differential equation.

Convergence

Intuitively, we may assume that a given consistent numerical scheme that remains stable,[1] produces numerical solutions that become closer to the exact solutions, as the time step tends toward 0. This property corresponds to the notion of convergence. In practice, only convergent schemes are used for simulating solutions.

Remark 3 For the special case of Eq. (3.2), we can add another property deduced from the knowledge of the exact solution. Indeed, we know that the solution must retain the sign of the initial condition. Let us say that $u^0 > 0$, thus $u_{\text{exact}}(t) > 0$ at any time. This property is called *positivity*.

At a discrete level, this is traduced by a condition for which $\dfrac{u^{k+1}}{u^k}$ remains positive. Thus, this may reduce the range for choosing a practical time step Δt for a given numerical scheme.

Exercise 4 Consider, once again, the scheme for Eq. (3.2) depending on the real parameter $\theta \in [0, 1]$:

$$\frac{u^{n+1} - u^n}{\Delta t} = -\lambda \left(\theta u^n + (1 - \theta) u^{n+1} \right) \qquad \theta\text{-scheme}$$

---

[1] Here it should be understood as zero-stability.

What is the positivity condition in the function of parameter θ?

$\qquad$ *Answer:* $\theta < \dfrac{1}{\lambda \, \Delta t}$.

Remark 4 The above partial results show that implicit schemes are better than explicit schemes, at least for the Euler approach. However, this may not be systematically true, as shown in the following illustration. Consider

$$u'(t) = u \qquad \text{with} \qquad u(0) = 1,$$

for which the exact solution is $u(t) = \exp(t)$. Thus, for a fixed time step Δt,

- the explicit Euler scheme reads as $u_k = (1 + \Delta t 1)^k$. Thus, $u_n = (1 + h)^n \rightarrow \exp(T)$, as $n \rightarrow +\infty$. The numerical solution seems to be acceptable even for *large* time steps.
- the implicit Euler scheme reads as $u_{k+1} = \dfrac{1}{1 - \Delta t} u_k$. First, note that this scheme is not defined for $\Delta t = 1$. Moreover, when Δt becomes closer to 1, the numerical solution explodes. For $\Delta t > 1$, the positivity of the solution is lost, and for $\Delta t = 2$, the numerical solution oscillates between $+1$ and -1.

This behavior would be better understood by formally defining A-stability.

3.1.2   Generalization

Consider the general expression of a first-order scalar ODE

$$\frac{du}{dt} = F(u, t), \tag{3.5}$$

where u is a scalar function depending on the time and F is any function, which is mostly nonlinear. As shown earlier, the key for discretizing the ODE is the approximation of the derivative; this can be done through several approaches by using the Taylor formula. In addition, we show the accuracy of the approximations, and the *regions of stability* of some popular schemes.

3.1.2.1   Taylor Expansion Formula

Recall that for any smooth enough function f, the Taylor expansion at any time t is

$$f(t + \Delta t) = f(t) + \Delta t \, f'(t) + \frac{\Delta t^2}{2} \, f''(t) + \cdots + \frac{\Delta t^n}{n!} \, f^{(n)}(t) + \mathcal{O}(\Delta t^{n+1}). \tag{3.6}$$

This formula shows that at any time t^n, it is easy to compute some approximation of the derivatives of a smooth function at any order. For example, for the first derivative,

$$u'(t^n) = \frac{u^{n+1} - u^n}{\Delta t} + \mathcal{O}(\Delta t),$$

$$u'(t^n) = \frac{u^n - u^{n-1}}{\Delta t} + \mathcal{O}(\Delta t),$$

$$u'(t^n) = \frac{u^{n+1} - u^{n-1}}{2\Delta t} + \mathcal{O}(\Delta t^2),$$

$$u'(t^n) = \frac{-u^{n+2} + 4u^{n+1} - 3u^n}{2\Delta t} + \mathcal{O}(\Delta t^2),$$

$$u'(t^n) = \frac{u^{n-2} - 4u^{n-1} + 3u^n}{2\Delta t} + \mathcal{O}(\Delta t^2).$$

The Taylor expansion formula in Eq. (3.6) is the fundamental tool of the finite difference approach.

Exercise 5 Show that the following approximations are accurate up to order $\mathcal{O}(\Delta t^4)$:

$$u'(t^n) \simeq \frac{u^{n-2} - 8u^{n-1} + 8u^{n+1} - u^{n+2}}{12\Delta t},$$

$$u''(t^n) \simeq \frac{-u^{n-2} + 16u^{n-1} - 30u^n + 16u^{n+1} - u^{n+2}}{12\Delta t^2}.$$

3.1.2.2   FE and BE Schemes Revisited

Consider the ODE evaluated at time t^n

$$\left(\frac{du}{dt}\right)\bigg|_{t^n} = F(u^n, t^n)$$

where $u^n \simeq u(t^n)$. There are (at least) two simple ways to approximate the derivative occurring at the LHS:

FE scheme: The approximation uses the next discrete point t^{n+1}, and the resulting scheme is given by the so-called FE scheme as

$$u^{n+1} = u^n + \Delta t \, F(u^n, t^n).$$

The unknown u^{n+1} can be computed by using an explicit formula. As a result, the FE scheme is called an explicit method.

BE scheme: The approximation uses the previous discrete point t^{n-1}, and the resulting scheme is given by

$$u^n = u^{n-1} + \Delta t \, F(u^n, t^n).$$

Up to a shift of the upperscript, this is equivalent to

$$u^{n+1} = u^n + \Delta t \, F(u^{n+1}, t^{n+1}),$$

which can be recast in the usual form as the BE scheme:

$$u^{n+1} - \Delta t \, F(u^{n+1}, t^{n+1}) = u^n.$$

The two terms of the LHS contain the unknown u^{n+1}, whereas the RHS is already given. Thus, this formulation shows that the unknown is not explicitly given; an inversion of the operator occurring at the LHS of the scheme is needed. Thus, the method is often called *implicit*.

3.1.2.3   Other Popular Schemes

The Euler schemes are certainly the most simple, and perhaps the most widely used for time-marching algorithms devoted to space and time simulations (cf. the next Sect. 3.2 on PDEs). They belong to the class of one-step algorithm, in which to compute a numerical solution at a given time, we must know only what happens, at most, one time step before. However, the counterpart of the simplicity of the BE and FE schemes is the lack of accuracy. To obtain higher accuracy, a higher order approximation of the differential operator must be determined using the Taylor expansion formula. This irremediably involves at least as numerous discrete variables as the order of accuracy of the approximation. The resulting scheme may thus contain a large number of elements occurring in the *stencil* scheme.

Another more convenient way to increase the accuracy for one-step methods is to compute intermediate values between t^n and $t^n + \Delta t$. The computation of intermediate values is accomplished at different stages of the time intervals, for which the approximation of the derivatives is accurate. This class of methods is called r-stage methods. The most famous r-stage algorithms are the RK methods.

The other class of methods allowing high-accuracy approximation of derivatives is the Linear Multi-steps Methods (LMM). Here, instead of computing values at intermediate stages within one time step, previously computed values are involved. Thus, no additional computation is needed. However, these methods need initial values when applied to initial value problems. These initial values must be computed using one-step methods.

RK type: They correspond to computational methods of u^{n+1} that involve intermediate stages between t^n and t^{n+1}, allowing to obtain a higher accuracy without
involving higher-order derivatives. The order of the RK methods corresponds to
the number of intermediate evaluations of function F occurring at the RHS of the
differential equation. These intermediate evaluations occur at different interme-

diate times (stages) between t^n and t^{n+1}. The general formulation of S-stage RK methods is as follows:

- Define S values as

$$k_1 = F(u, t),$$
$$k_2 = F(u + \beta_{2,1} k_1 \Delta t, \ t + \alpha_1 \Delta t),$$
$$\vdots$$
$$k_S = F\left(u + \sum_{i=1}^{S-1} \beta_{S,i} k_i \Delta t, \ t + \alpha_S \Delta t\right).$$

The coefficients $\beta_{i,j}$ correspond to the weights of each intermediate function evaluations, and α_i are related to the intermediate times.
- The S-stage RK methods are then defined by

$$u^{n+1} = u^n + \Delta t \sum_{i=1}^{S} \gamma_i k_i.$$

By fixing all the coefficients α_i, $\beta_{i,j}$, and γ_i for all $i = 1, \ldots, S$ stages, a particular RK scheme is defined, with specific properties about factors such as accuracy and stability.[2] The computation of these coefficients follows some given rules, such as $\sum_i \gamma_i = 1$. In addition, it depends on the accuracy that must be achieved. Thus, the choice of all the weights of coefficients is not arbitrary.

Some examples of particular cases are as follows.

- RK 1: Here only one stage interval k_1 is considered for computing u^{n+1} from u^n as

$$u^{n+1} = u^n + \Delta t F(u^n, t^n).$$

This method is analogous to the FE scheme.
- RK 2 methods read as

$$u^{n+1} = u^n + \gamma_1 \Delta t \, F(u^n, t^n) + \gamma_2 \Delta t \, F\left((u + \beta_{2,1} k_1 \Delta t, \ t + \alpha_1 \Delta t\right).$$

  - By taking the particular values $\gamma_1 = \gamma_2 = \frac{1}{2}$, and $\alpha_1 = \beta_{2,1} = 1$, we obtain the so-called Heun method as follows:

$$u^{n+1} = u^n + \frac{\Delta t}{2} F(u^n, t^n) + \frac{\Delta t}{2} F\left((u^n + \Delta t F(u^n, t^n), \ t^n + \Delta t\right).$$

---

[2]The development of such methods is also devoted to the preservation of the symplecticity of the scheme, as it is highly recommended in celestial mechanics problems such as N-body interactions.

– By taking the particular values $\gamma_1 = \gamma_2 = \frac{1}{2}$, and $\alpha_1 = \beta_{2,1} = \frac{1}{2}$, we obtain the so-called Predictor—Corrector method as follows:

$$u^{n+1} = u^n + \Delta t\, F(u^n, t^n) + \Delta t\, F\left((u^n + \frac{\Delta t}{2} F(u^n, t^n),\ t^n + \frac{\Delta t}{2} \right).$$

• RK 4: Consider the four intermediate computations of F: $k_1, k_2, k_3,$ and k_4, as shown in Fig. 3.3.

$$k_1 = F(u^n, t^n), \qquad\qquad\qquad k_2 = F(u^n + k_1\frac{\Delta t}{2},\ t^n + \frac{\Delta t}{2}),$$

$$k_3 = F(u^n + k_2\frac{\Delta t}{2},\ t^n + \frac{\Delta t}{2}), \qquad k_4 = F(u^n + k_3\Delta t,\ t^n + \Delta t).$$

Then, a widely used RK 4 method reads as

$$u^{n+1} = u^n + \frac{\Delta t}{6}\, (k_1 + 2k_2 + 2k_3 + k_4).$$

The order of accuracy of the RK methods is given by their number of stages (up to the number of stages equal to four, later the relation between the accuracy order and the number of stages becomes much more complicated and little is known).

Linear multi-step methods: They correspond to computational methods of u^{n+r} not involving the knowledge of only u^{n+r-1}; however, eventually all the previous nodes from u^n to u^{n+r-1} are known, that is,

$$\sum_{j=0}^{r} \alpha_r u^{n+r} = \Delta t \sum_{j=0}^{r} \beta_j F(u^{n+j}).$$

Among the most popular, the general Adams method is as follows:

$$u^{n+r} = u^{n+r-1} + \Delta t \sum_{j=0}^{r} \beta_j F(u^{n+j}).$$

The coefficients β_j are computed in the function of the expected accuracy. Moreover, for computing u^{n+r}, only u^{n+r-1} is directly used, as the other previous values are involved through the RHS function F.

• Adams–Bashforth method: Here, the coefficient $\beta_r = 0$. Moreover, β_j, $j = 0, \ldots, r-1$ is taken such that the method is accurate to the order r. These methods require r steps and are explicit methods. One of the most widely used methods is

$$u^{n+1} = u^n + \frac{3}{2}\Delta t\, F(u^n, t^n) - \frac{1}{2}\Delta t\, F(u^{n-1}, t^{n-1}).$$

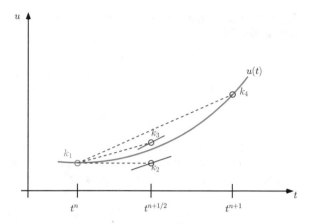

Fig. 3.3 Intermediate computations of RK4 method. Segments of same color have identical slopes

- Adams–Moulton method is the implicit counterpart of the Adams–Bashforth method. The coefficient β_r is chosen as not null, yielding an r-step accuracy of order $r + 1$. One of the most famous methods is the midpoint method, with the trapezoidal rule:

$$u^{n+1} = u^n + \frac{1}{2} \Delta t \left(F(u^{n+1}, t^{n+1}) + F(u^n, t^n) \right).$$

The stability of such linear multi-step methods is a notion to be defined precisely. Moreover, the stability analysis requires the introduction of new tools. These concepts give rise to important issues.

3.1.2.4 Accuracy, Consistency, Stability, and Convergence

Accuracy: When a derivative is approximated using a finite number of discrete unknowns, the computed numerical solution corresponds to an approximation of the exact solution up to the error induced by the neglected terms. These neglected terms, also called local truncation errors, depend on the time step used and are proportional to the higher-order derivatives. For Euler methods, a higher order of truncation in the Taylor development implies a more accurate approximation of the derivative.

Formally, the error of truncation can be defined as follows. Let us note the differential operator $\mathscr{L}$ associated to the differential equation $\mathscr{L}(u) = 0$, with the exact solution: u_{exact}. Assume an approximation of $\mathscr{L}$ computed using the Taylor expansion $\mathscr{L}_{\Delta t}$. Thus, the associated finite-difference scheme is $\mathscr{L}_{\Delta t}(u) = 0$. The local truncation error ε is then defined as follows:

$$\varepsilon_{\Delta t} := \mathscr{L}_{\Delta t}(u_{\text{exact}}).$$

Exercise 6 Find the order of the local truncation error $\varepsilon_{\Delta t}$ associated with the leap-frog method:

$$\frac{u^{n+1} - u^{n-1}}{2\Delta t} = F(u^n). \tag{3.7}$$

Answer: $\varepsilon_{\Delta t} = \dfrac{\Delta t^2}{12} u''' (t^n) + \cdots \simeq \mathcal{O}(\Delta t^2)$

Consistency: Another important notion to be introduced here is the global error $\mathscr{E}^n$ obtained by a numerical solution. It corresponds to the difference, at a given time t^n, between the exact and numerical solutions:

$$\mathscr{E}^n = \left| u^n - u_{\text{exact}}(t^0 + n\Delta t) \right|.$$

As u^n is computed using a numerical scheme of a given order of accuracy, the difference between the exact and numerical solutions is related to the accuracy of the method. Thus, there is a relation between the global error $\mathscr{E}^n$ (which compares the disparity of a numerical solution from the exact solution) and the local truncation error $\varepsilon_{\Delta t}$ (which shows the accuracy of the used method). The notion of consistency can be formally written as follows:

$$\forall n \quad \mathscr{E}^n \to 0 \qquad \text{when} \qquad \Delta t \to 0.$$

Exercise 7 Show that the leap-frog method in Eq. (3.7) is consistent with Eq. (3.5).

Sometimes $\mathscr{E}$ is called the tolerance, or the acceptable precision, on the computed solution.

Remark 5 Here, we provide the first answer to one of the main problems given in the general introduction in 3.1. Suppose we want to simulate a dynamical system over a period T. The expected numerical solution u^N, with $N\Delta t = T$ must have an error of order $\mathscr{E} \simeq 10^{-\eta}$. If a numerical method accurate up to order $\mathcal{O}(\Delta t^k)$ is used, then it is expected to run the algorithm $N = T \times 10^{\eta/k}$.

For example, if the tolerance on the solution is $\left| u^N - u_{\text{exact}}(T) \right| \simeq 10^{-4}$, and a first-order accuracy method is used, the number of time iterations would be of order $N_1 \simeq T \times 10^4$, whereas if a fourth-order accuracy method is used, then the number of time iterations would be $N_4 \simeq T \times 10$.

Stability: An intuitive idea of the stability for an ODE can be conceptualized by assuming (any) two simulations with two closed initial conditions. It is expected that for a given fixed time, the solutions can be found as closed as desired. If this is false, the difference between the solutions remains unbounded, regardless of the choice of initial conditions.[3] Theoretically, a *zero-stability* for schemes can be formally given by

---

[3] Recall that we are focusing on the solutions produced by a numerical algorithm and not the solutions of a continuous model.

$$\mathcal{E}^n \to 0 \qquad \text{when} \qquad \Delta t \to 0, \ n \to \infty.$$

It can be proved that all one-step methods have zero-stability.

Exercise 8 Show that the scheme in Eq. (3.7) is of zero-stability.

In practice, Δt is never null. It is often a constant (similar to the methods presented here), or it may change during the simulation (time adaptive schemes), and thus the notion of *zero-stability* is not very useful. In practice, it is more relevant to consider *A-stability* of a numerical scheme. *A-stability* (A as absolute) is commonly described using the complex plan $\mathbb{C}^4$ as follows. A region of *A-stability* is computed by defining the values of the time step Δt for which the numerical scheme does not give spurious solutions. The techniques used for computing such a region of *A-stability* are not described here. However, we provide the results about some popular schemes. Two cases must be described corresponding to whether the equation is linear.

1. Linear case: $F(u, t) = a(t)u(t) + b(t)$.

   - Euler schemes: The stability analysis follows the same main steps as those for the exponential equation. For the FE scheme,

   $$u^{n+1} = u^n + \Delta t\, F(u^n, t^n),$$

   and BE scheme,

   $$u^{n+1} = u^n + \Delta t\, F(u^{n+1}, t^{n+1}),$$

   one has to recast the terms of a geometric sequence as done earlier:

   $$u^{k+1} = q\, u^k, \ k \in \mathbb{N}$$

   The condition of stability corresponds to the condition for $|q| \leq 1$.

2. Nonlinear case: there is no systematic tool to exactly describe the stability region of a given method applied to a nonlinear equation. We can just obtain an estimation by linearizing the equation (in practice, this is often a good approximation as long as the solution does not vary too rapidly for the time step used). Set $v(t) = u(t) - u^\star$, then,

$$
\begin{aligned}
v'(t) &= u'(t) \\
&= f(u(t), t) \\
&= f(v(t) + u^\star, t),
\end{aligned}
$$

---

[4]Consider the exponential equation $u' = -\lambda u$, and assume that λ can be complex. Thus, the stability condition, as computed previously for Euler schemes, must be described following both real and imaginary parts of λ.

and a Taylor expansion around $u^\star$ gives $v'(t) = f(u^\star, t) + v(t) f'(u^\star, t) + \mathcal{O}(v^2)$. Thus, v is the solution of the linear system:

$$v'(t) = Av(t) + b,$$

with $A = f'(u^\star, t)$ and $b = f(u^\star, t)$. Furthermore, the A-stability of the linear equation for relevant value of $u^\star$ must be analyzed.

Convergence: Intuitively, a scheme is convergent when its solutions tends toward the exact solution as the time step tends toward zero, and when the solutions remain well behaved. It corresponds to the properties of consistency added to (zero-) stability. However, as the convergence of a numerical scheme is, in practice, an absolute necessity, its A-stability region must be focused upon.

Exercise 9 Can the scheme in Eq. (3.7) be used?
Answer: No, because its stability condition is reduced to 0, as observed at the axis of the real numbers.

Now, we can answer the main question stated in the Introduction (Sect. 3.1). In practice, the A-stability region defines the bounds within which the time step Δt has to be considered. It depends on the numerical scheme and on the equation (cf. Remark 4). Let Δt_s be the biggest time step authorized by using an A-stability analysis. Let Δt_a be the biggest time step that can be chosen such that the tolerance on the numerical solution is satisfied (cf. the practical Remark 5). Thus,

1. if $\Delta t_s > \Delta t_a$, then Δt must be taken equal to Δt_a.
2. if $\Delta t_s < \Delta t_a$, then the numerical method must probably be changed for a more relevant method.

3.1.3 Systems of ODEs

This section shows our exploitation of the scalar case study to extend the analysis and apply numerical schemes to a system of ODEs. The cases of higher-order linear differential equation and explicitly time-dependent system are used again to study an autonomous system of ODE. Thus, the numerical schemes viewed previously can be applied formally and directly to systems of ODEs, whereas the analysis of the numerical properties is slightly more complex than the scalar case. Finally, two common families of system of ODE are very briefly presented, along with an illustration that in real life problems, mathematical models may need sophisticated and technical methods to be well simulated.

3.1.3.1 Higher-Order Linear Differential Equation, Non-autonomous Case

Higher-Order Linear Differential Equation

Higher-order linear scalar ODE can be recast into a system of first-order ODEs. Consider a differential equation involving at most an N-order derivative:

$$\alpha_N \frac{d^{(N)}u}{dt^N} + \cdots + \alpha_k \frac{d^{(k)}u}{dt^k} + \cdots + \alpha_1 \frac{du}{dt} + \alpha_0 u = 0.$$

Then, it can be recast into a system of N first-order ODEs as

$$\frac{d\mathbf{u}}{dt} = \mathscr{A}\mathbf{u},$$

where $\mathbf{u}(t) \in \mathbb{R}^N$ and $\mathscr{A} : \mathbb{R}^N \to \mathbb{R}^N$ gives a linear mapping by changing variables as

$$\frac{d^{(N-1)}u}{dt^{N-1}} = u^{N-1},$$

$$\vdots$$

$$\frac{d^{(k)}u}{dt^k} = u^k,$$

$$\vdots$$

$$\frac{du}{dt} = u^1,$$

and setting the new vector of variables $\mathbf{u} = \{u^N, \ldots, u^{k+1}, \ldots, u^2, u\} \in \mathbb{R}^N$. For example, consider simply the case for a second-order ODE, $N = 2$:

$$\alpha_2 \frac{d^2 u}{dt^2} + \alpha_1 \frac{du}{dt} + \alpha_0 u = 0.$$

Then, the associated first-order matrix system is as follows:

$$\frac{d}{dt}\begin{pmatrix} u_1 \\ u \end{pmatrix} = \begin{pmatrix} -\frac{\alpha_1}{\alpha_2} & -\frac{\alpha_0}{\alpha_2} \\ 1 & 0 \end{pmatrix}\begin{pmatrix} u_1 \\ u \end{pmatrix}.$$

Non-autonomous Case

When the RHS depends explicitly on time t, it is possible to recast a system of dimension d, $\bar{u} \in \mathbb{R}^d$ into a time independent system of dimension $d + 1$ to construct an equivalent autonomous system. For this, add the $(d+1)$th component corresponding to the time $u_{d+1} = t$, and set $\dfrac{du_{d+1}}{dt} = 1$.

3.1.3.2   Natural Generalization of the Scalar Case

Numerical Schemes

All the methods presented previously, for the scalar case, can be transposed to the multidimensional case, except instead of considering $u \in \mathbb{R}$, consider $\mathbf{u} \in \mathbb{R}^N$. For example, for the Euler methods using the matrix formalism, the system of first-order ODE is as follows.

- Forward time-centered space scheme (FTCS)

$$\mathbf{u}^{n+1} = \mathbf{u}^n + \Delta t \, \mathscr{A}(\mathbf{u}^n).$$

- Backward time-centered space scheme (BTCS)

$$\mathbf{u}^{n+1} = \mathbf{u}^n + \Delta t \, \mathscr{A}(\mathbf{u}^{n+1}).$$

- θ-scheme

$$\mathbf{u}^{n+1} = \mathbf{u}^n + \Delta t \left[\theta \, \mathscr{A}(\mathbf{u}^{n+1}) + (1 - \theta) \, \mathscr{A}(\mathbf{u}^n) \right].$$

The special case $\theta = 1/2$ is called a trapezoidal rule scheme.

Accuracy, Stability, and Consistency

The notions of accuracy and consistency (and thus error) are the same as in the scalar case. Here, only the notion of A-stability may have to be precise.

- Consider the linear case, $F(\mathbf{u}) = \mathscr{A}\mathbf{u}$, $\mathbf{u} \in \mathbb{R}^N$. Assume $\mathscr{A}$ is diagonalizable: $\mathscr{A} = P \mathscr{D} P^{-1}$. Then, the variable $\mathbf{z} = P^{-1}\mathbf{u} \in \mathbb{R}^N$ satisfies the diagonal-matrix system

$$\frac{d\mathbf{z}}{dt} = \mathscr{D}\mathbf{z},$$

that corresponds to N uncoupled linear equations. Now, we only need to apply the results previously exposed for the scalar case to study the system. Finally, this approach corresponds to describe the spectrum of the linear application $\mathscr{A} : \mathbb{R}^N \to \mathbb{R}^N$, that is, to determine its eigenvalues $\lambda_1, \ldots, \lambda_N$.
- Similar to the scalar case, for the nonlinear situation $F(\mathbf{u}) = \mathscr{A}(\mathbf{u})$, we must linearize the system and consider the Jacobian matrix $J_{\mathscr{A}}$ as

$$J_{\mathscr{A}}(\mathbf{u}) = \begin{pmatrix} \dfrac{\partial A_1}{\partial u_1} & \cdots & \dfrac{\partial A_1}{\partial u_N} \\ & \vdots & \\ \dfrac{\partial A_N}{\partial u_1} & \cdots & \dfrac{\partial A_N}{\partial u_N} \end{pmatrix}$$

Thus, a similar change of variables (as for the scalar case) must be set as $\mathbf{v}(t) = \mathbf{u}(t) - \mathbf{u}^\star$ for solution of the linear system.

$$\frac{d\mathbf{v}}{dt} = J_{\mathscr{A}}(\mathbf{u}^\star)\mathbf{v} + \mathscr{A}(\mathbf{u}^\star).$$

3.1.3.3 Chaotic and Stiff Problems

It is possible that standard schemes cannot succeed in efficiently simulating the behavior of some system of ODEs. This may be due to the *chaotic* nature of the system, that is, a slight perturbation of the initial condition induces a relatively different quantitative (and sometimes qualitative) behavior of the solution. This behavior is a consequence of high nonlinear effects. A well-known illustration of this behavior is given through meteorological dynamics, in which local perturbations (butterfly effects) may induce severe global changes. Another common reason for the failure of standard numerical methods in efficiently simulating the dynamics of a system is due to intrinsic different time scales. Intuitive illustrations can be obtained through celestial configurations, in which, for example, the time scale related to planetary orbits are many orders of magnitude smaller than those of galaxies. Thus, from the numerical viewpoint, adapted techniques are needed. Let us illustrate two common cases occurring in Engineering Science.

Chaotic systems: These occur in Ecology, Biology, and even in Economy. The most famous example is given by the Lorenz system (occurring in fields such as chemical engineering, electrical engineering, and meteorology):

$$\begin{cases} x'(t) = \sigma\,(y(t) - x(t)), \\ y'(t) = x(t)\,(\rho - z(t)) - y(t), \\ z'(t) = x(t)y(t) - \beta z(t), \end{cases}$$

where the parameters σ, ρ, and β take positive real values. For practical implementation of this system, time adaptive schemes are needed.

Stiff systems: These occur in Chemistry, for example. A formal criterion for the stiffness of a given system of ODE is

$$\frac{d\mathbf{u}}{dt} = \mathscr{A}(\mathbf{u}).$$

It is given by the ratio of the extremum eigenvalues of the Jacobian $J_{\mathscr{A}}$:

$$\mathscr{S}(J_{\mathscr{A}}) = \frac{\max |\lambda_i|}{\min |\lambda_i|}.$$

When this ratio is very big, that is, $\mathscr{S} \gg 1$, the system may present a time interval for which time scales of a very different order of magnitude are involved, making the dynamics of the system very difficult to capture.

3.1.4 Exercises

3.1.4.1 Theoretical Part

Highest Time Step for a Given Tolerance

Assume a numerical method accurate up to the fourth order, without any stability restrictions. We require to run a simulation over a time $T = 1$, with, an acceptable numerical solution up to a tolerance 10^{-8}.

1. What is the highest time step that can be used *a priori*?
2. What is the number of iterations for the simulation?
3. How can the bound time step be *theoretically* improved? Comment.

A-Stability of Euler Schemes for Linear System of ODE

Consider Euler methods for a linear system of ODE

$$\frac{d\mathbf{u}}{dt} = \mathscr{A}\mathbf{u},$$

Show that the stability condition is related to the positivity of the real part of each λ_i.

- If $\mathrm{Re}(\lambda_i) \leqslant 0, \forall \lambda_i$, then the scheme is conditionally stable for the FE scheme, and unconditionally stable for the BE scheme. The conditions of stability for the FE scheme are given by the following relations:

$$|1 + \lambda_i \Delta t| \leqslant 1, \qquad\qquad \text{for all}\quad i,$$

- If $\exists \lambda_j$ such that $\mathrm{Re}(\lambda_j) > 0$, then the scheme is unconditionally unstable for the FE scheme, and conditionally stable for the BE scheme. The conditions of stability for the BE scheme are given by the following relations:

$$\frac{1}{|1 - \lambda_j \Delta t|} \leqslant 1, \qquad\qquad \text{for all}\quad j \mid \mathrm{Re}(\lambda_j) > 0.$$

Numerical Conditioning

Consider the problem

$$u'(t) = 3t - 3u(t) \qquad \text{and} \qquad u(0) = \alpha \in \mathbb{R}$$

1. Show that $u(t) = (\alpha - \frac{1}{3}) \exp(3t) + t + \frac{1}{3}$ is the solution of the problem.
2. Compute $u(t = 10)$ with $\alpha = 1/3$.
3. Compute $u(t = 10)$ with $\alpha = 0.333333$.
4. What is the order of the error on α in both cases, and what is the order of the difference between the computed solution $u(10)$?

Application: The Harmonic Oscillator Toward Symplectic Schemes

Consider once again the harmonic 1D oscillator (i.e., the simple pendulum model):

$$\ddot{x} + \omega^2 x = 0,$$

where the pulsation ω is a function of the stiffness of the spring and mass. Recall that in the absence of the damping term, the mechanical energy $E = E_k + E_p$ (where the kinetic energy is $E_k = \frac{1}{2}m\dot{x}^2$ and the potential energy is $E_p = \frac{1}{2}kx^2$) is conserved along the time. This implies $\dot{E} = 0$ for all solutions of the system.

1. Recast the second-order ODE into a system of first-order ODEs

$$\dot{z} = Az, \qquad (3.8)$$

with

$$A = \begin{pmatrix} 0 & 1 \\ -\omega^2 & 0 \end{pmatrix} \qquad z = \begin{pmatrix} x \\ y \end{pmatrix} \qquad (3.9)$$

What is the physical interpretation of the variables x and y?
2. What is the discrete counterpart of the mechanical energy E^n, evaluated at time t^n in function (x^n, y^n)?
3. Write down explicit and implicit Euler schemes for the matrix system.
4. Hereafter, take $k = 1$ and $m = 1$. What is the numerical error induced by these schemes on the conservation of the discrete mechanical energy?
5. Let us intensively understand what transpires with this harmonic oscillator. Show that $(x(t_0 + \tau), y(t_0 + \tau))$ defined for any time t_0 as

$$\begin{pmatrix} x(t_0 + \tau) \\ y(t_0 + \tau) \end{pmatrix} = \begin{pmatrix} \cos(\tau) & \sin(\tau) \\ -\sin(\tau) & \cos(\tau) \end{pmatrix} \begin{pmatrix} x(t_0) \\ y(t_0) \end{pmatrix} \qquad (3.10)$$

is the solution of the system in Eq. (3.8). This implies that, for a given surface in the *phase space* representation defined by a set of initial conditions $\mathcal{D}_0 = \{x_0, y_0\}$, the surface evolves as $\mathcal{D}(\tau) = \{x(\tau), y(\tau)\}$, as defined by Eq. (3.10).

6. Recall that a change of volume can be traduced by the change of variables formula
   as

$$\int_{\mathscr{D}_0} \cdot \, dz = \int_{\mathscr{D}(\tau)} \cdot \mid \det J \mid dz$$

where J is the Jacobian of the transformation.

7. What is the value of $\mid \det J \mid$ for the exact solution? Interpret it on $\mathscr{D}(\tau)$.
8. What is the value of $\mid \det J \mid$ for the Euler schemes? Define the conclusion for $\mathscr{D}^n$.
9. Consider the following *symplectic* scheme:

$$\begin{pmatrix} x(t_0 + \tau) \\ y(t_0 + \tau) \end{pmatrix} = \begin{pmatrix} 1 & \tau \\ -\tau & 1 - \tau^2 \end{pmatrix} \begin{pmatrix} x(t_0) \\ y(t_0) \end{pmatrix} \tag{3.11}$$

What is the value of $\mid \det J \mid$ for this scheme? Provide the conclusion for $\mathscr{D}^n$.
10. Is the discrete mechanical energy conserved by this scheme?
11. Comment about the modified Energy $H = \frac{x^2 + y^2}{2} + \Delta t \frac{xy}{2}$?

3.2 Parabolic PDE

Recall that most classical models expressed as a PDE, occurring in engineering science, can be classified as follows according to the types of boundary conditions that must be solved[5]:

- Elliptic equation examples: the Laplace, Poisson, and Helmholtz equations, and more generally most of the steady-state models[6]
- Hyperbolic equations examples: pure advection equations, Euler equations for gas dynamics and fluid flows, and more generally the equations for which the velocity of transported information is finite[7]
- Parabolic equations: the case we are interested in, that is, the case in which some non-negligible diffusion processes occur; typically, the heat diffusion equation.

The present section shows the basic and popular schemes used for simulating a diffusion process. Their construction and algorithm to solve the resulting matrix system are detailed. Some applications in building physics engineering are presented.

References regarding PDEs are numerous. Let us instead cite the following reference books that treat the solution of diffusion process using methods that considerably surpass the finite difference approach

---

[5] More precisely, to be well-posed.

[6] Time-independent boundary value problems would be better.

[7] This has a mathematical meaning related to the considered differential operator.

- Ozisik, M. N., *Heat Conduction.* 1993. Wiley [150]
  A complete reference of analysis, techniques, and resolutions about transfer phe-
  nomena. Although it has been written many years ago, it is still relevant.
- Saatdjian, E., *Transport Phenomena: Equations and Numerical Solutions.* 2000.
  Wiley [169]
  A very comprehensive, practical, and well detailed source. Undoubtedly useful
  for engineering purposes.

3.2.1 The Heat Equation in 1D

The 1D heat equation is as follows:

$$\frac{\partial u}{\partial t} = \frac{\partial}{\partial x}\left(\alpha(u)\frac{\partial u}{\partial x}\right) + g(x, t),$$

where $u(x, t)$ is the diffusive scalar field, x is the space coordinate in the interval
$]a, b[$, and t represents the time coordinate. In the most general case, the diffusion
coefficient α may depend on u, or eventually only on x. The source term g must
expected to be known.

Hereafter, the general case is treated step-by-step; however, for now, consider the
simple case of a source of a less constant property diffusion equation:

$$\frac{\partial u}{\partial t} = \alpha\frac{\partial^2 u}{\partial x^2}. \tag{3.12}$$

3.2.1.1 Euler Schemes

The first operation for computing numerical schemes for solving the heat equation
involves the discretization of both time and space lines so that we obtain a time–space
grid. Thus, Δt and Δx are the time and space steps such that,

$$t^n = t^0 + n\Delta t, \qquad \forall n \in \mathbb{N}$$
$$x_j = a + j\Delta x, \qquad \forall j = 0, 1, 2, \ldots, N.$$

Remark 6 We could have chosen an irregular discretization of the independent variable so that the step size would depend on the point discretization. For the time
variable, this yields a time-adaptive process, whereas for a space variable, this corresponds to refined mesh techniques (see Fig. 3.4).

Exercise 10 Suppose the space interval length is $L = b - a$, and the discretization
of this segment involves exactly $N + 1$ points regularly spaced. What is the expression of Δx?

Fig. 3.4 Cartesian grid in
one space and time
dimension

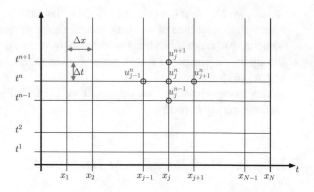

Answer: $\Delta x = \dfrac{b - a}{N}$

Next, for ODE, each differential operator must be approximated using the Taylor expansion. Let us begin with the time variable so that the semi-discretized heat equation becomes

$$\frac{u^{n+1}(x) - u^n(x)}{\Delta t} = \alpha \frac{\partial^2 u^n(x)}{\partial x^2}, \qquad \text{FE Scheme in Time}$$

$$\frac{u^{n+1}(x) - u^n(x)}{\Delta t} = \alpha \frac{\partial^2 u^{n+1}(x)}{\partial x^2}, \qquad \text{BE Scheme in Time}$$

where $u^n(x)$ is the approximation of the heat equation solution $u(t^n, x)$.

Now, by using a centered approximation for the second derivative (diffusion term),

$$f''(x) = \frac{f(x - \Delta x) - 2f(x) + f(x + \Delta x)}{\Delta x^2} + \mathcal{O}(\Delta x^2).$$

Finally, we obtain the finite differences approximations as

$$\frac{u_j^{n+1} - u_j^n}{\Delta t} = \alpha \frac{u_{j-1}^n - 2u_j^n + u_{j+1}^n}{\Delta x^2}, \qquad \text{FE in Time, Centered Space Discretization}$$

$$\frac{u_j^{n+1} - u_j^n}{\Delta t} = \alpha \frac{u_{j-1}^{n+1} - 2u_j^{n+1} + u_{j+1}^{n+1}}{\Delta x^2}, \qquad \text{BE in Time, Centered Space Discretization}$$

where u_j^n represents the numerical approximation of $u(t^n, x_j)$ solution of Eq. (3.12). A more convenient formulation for computational purpose reads as follows (Fig. 3.5):

- FTCS:

$$u^{n+1} = (1 - 2\mathsf{Fo})\, u_j^n + \mathsf{Fo}\left(u_{j-1}^n + u_{j+1}^n\right).$$

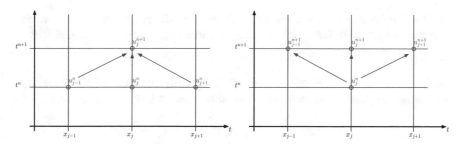

Fig. 3.5 The left stencil represents the FTCS, and the right stencil represents the BTCS

- BTCS:

$$-\mathrm{Fo}\, u_{j-1}^{n+1} + (1 + 2\mathrm{Fo})\, u_j^{n+1} - \mathrm{Fo}\, u_{j+1}^{n+1} = u_j^n,$$

where the Fourier number $\mathrm{Fo} \overset{\text{def}}{:=} \dfrac{\alpha\,\Delta t}{\Delta x^2}$ appears.[8]

Remark 7 To understand a finite differences scheme, consider a function u that verifies a given differential operator $\mathscr{L}(u) = 0$. A finite differences scheme is an algebraic operator $\mathscr{L}_h$, operating on a set of point-values of functions. Thus, the differential operator $\mathscr{L}$, operating on $\mathscr{D} = \{$set of functions regular enough$\}$ is substituted by an algebraic operator $\mathscr{L}_h$, operating on a Cartesian product of $\mathscr{D}$. This implies that the numerical solution is sought with the same regularity as in the continuous solution. The consistency and stability correspond to $\mathscr{L}_h$ being a bounded operator that tends toward $\mathscr{L}$ when $h \to 0$.

Accuracy

Both schemes are accurate up to order $\mathscr{O}(\Delta t + \Delta x^2)$. Indeed, for each time–space point (t^n, x_j), we have exactly

$$\frac{u(t^{n+1}, x_j) - u(t^n, x_j)}{\Delta t} - \alpha\,\frac{u(t^n, x_{j-1}) - 2u(t^n, x_j) + u(t^n, x_{j+1})}{\Delta x^2}$$
$$= \mathscr{O}(\Delta t + \Delta x^2), \quad \text{FTCS}$$

$$\frac{u(t^{n+1}, x_j) - u(t^n, x_j)}{\Delta t} - \alpha\,\frac{u(t^{n+1}, x_{j-1}) - 2u(t^{n+1}, x_j) + u(t^{n+1}, x_{j+1})}{\Delta x^2}$$
$$= \mathscr{O}(\Delta t + \Delta x^2), \quad \text{BTCS}$$

---

[8]The name of Fourier is relevant when considering α as a thermal diffusion coefficient. In computational fluid dynamics, α represents viscosity and is rather called the CFL number, thanks to Courant, Friedrichs, and Lewy who in 1928 discovered a stability condition for numerical schemes involving this term.

where u is solution of the heat equation. Thus, we easily recognize those numerical schemes on each LHS for which solutions are approximations of u modulo RHS.

Stability

The standard way to perform a von Neumann stability analysis is by considering a β-mode, corresponding to the Fourier decomposition of heat equation solutions:

$$u_j^n \equiv \rho^n e^{2 i \pi \beta x_j} \qquad \text{with} \qquad x_j = j \, \Delta x,$$

where ρ is the amplitude of the mode.

Note that for $\rho = \dfrac{u^{n+1}}{u^n}$, it is expected that for a free diffusion process, this ratio never becomes greater than 1 in the absolute value.[9] Otherwise, the numerical solution may gradually increase with n, thus yielding a divergent solution. Therefore, $|\rho| \leqslant 1$ is taken as the stability condition for numerical schemes associated to a diffusion process. By injecting the mode $\rho^n e^{2 i \pi \beta x_j}$, regardless of the mode number β, into the FTCS scheme, we get

$$\rho = (1 - 2\mathsf{Fo}) + \mathsf{Fo} \left(e^{2 i \pi \beta \, \Delta x} + e^{-2 i \pi \beta \, \Delta x} \right).$$

Recall that $\cos(\xi) = \dfrac{e^{i\xi} + e^{-i\xi}}{2}$ and $2 \sin^2(\xi) = 1 - \cos(2 \xi)$, so that the stability condition $|\rho| \leqslant 1$ reads as follows:

$$\left| 1 - 4\mathsf{Fo} \sin^2(\pi \beta \Delta x) \right| \leqslant 1, \qquad \forall \beta,$$

that is to say:

$$\mathsf{Fo} = \frac{\alpha \Delta t}{\Delta x^2} \leqslant \frac{1}{2} \qquad \text{Stability condition for FTCS scheme.}$$

This stability condition for the FTCS scheme is called the CFL condition, named after Courant, Friedrichs, and Lewy, who discovered this formula before the development of the first computer. A similar computation for the BTCS scheme shows that the associated stability condition reads as

$$\left| \frac{1}{1 + 4\mathsf{Fo} \sin^2(\pi \beta \Delta x)} \right| \leqslant 1, \qquad \forall \beta.$$

This equality is always verified, regardless of the values of the time and space steps. Thus, the BTCS scheme is said to be unconditionally stable.

---

[9]This expectation is related to some maximum principle, according to which when no source term exists, the diffusive field must decrease in magnitude as time passes.

3.2.1.2 Other Standard Schemes

We could construct an infinite number of schemes, involving different *stencils*, that is, nodes, of the mesh discretization occurring in the numerical scheme. All of these have different properties in accuracy, stability, and in the easy-to-implement computational cost (for 1D heat equation, this should not be a major problem). Hereafter, some of the most well-known schemes for the heat equation are provided (Fig. 3.6):

θ-scheme: For $\theta \in [0, \ 1]$,

$$\frac{u_j^{n+1} - u_j^n}{\Delta t} = \alpha \left[\theta \left(\frac{u_{j-1}^{n+1} - 2u_j^{n+1} + u_{j+1}^{n+1}}{\Delta x^2} \right) + (1 - \theta) \left(\frac{u_{j-1}^n - 2u_j^n + u_{j+1}^n}{\Delta x^2} \right) \right].$$

The FTCS scheme is a special case corresponding to $\theta = 0$, and the BTCS scheme corresponds to $\theta = 1$.

Crank–Nicolson scheme: This is a particular case of the θ scheme in which $\theta = 1/2$. It is perhaps the most widely used schemes for 1D heat equations:

$$\frac{u_j^{n+1} - u_j^n}{\Delta t} = \alpha \left[\left(\frac{u_{j-1}^{n+1} - 2u_j^{n+1} + u_{j+1}^{n+1}}{2\Delta x^2} \right) + \left(\frac{u_{j-1}^n - 2u_j^n + u_{j+1}^n}{2\Delta x^2} \right) \right].$$

Gear scheme:

$$\frac{3u_j^{n+1} - 4u_j^n + u_j^{n-1}}{2\Delta t} = \alpha \frac{u_{j-1}^{n+1} - 2u_j^{n+1} + u_{j+1}^{n+1}}{\Delta x^2}.$$

DuFort–Frankel scheme:

$$\frac{u_j^{n+1} - u_j^{n-1}}{2\Delta t} = \alpha \frac{u_{j-1}^n - u_j^{n+1} - u_j^{n-1} + u_{j+1}^n}{\Delta x^2}.$$

Exercise 11 For all the above-mentioned schemes, determine the order of accuracy and their stability condition.

Answer:

Scheme	Local error	Stability condition
θ-scheme, $\theta \neq \frac{1}{2}$	$\mathcal{O}(\Delta t + \Delta x^2)$	if $\Delta t \leqslant \dfrac{\Delta x^2}{2\alpha}$
Crank–Nicolson	$\mathcal{O}(\Delta t^2 + \Delta x^2)$	Stable
Gear	$\mathcal{O}(\Delta t^2 + \Delta x^2)$	Stable
DuFort–Frankel	$\mathcal{O}(\dfrac{\Delta t^2}{\Delta x^2} + \Delta x^2)$	if $\dfrac{\Delta t}{\Delta x^2}$ remains bounded

Fig. 3.6 Stencils for
Crank–Nicolson, Gear, and
DuFort–Frankel schemes

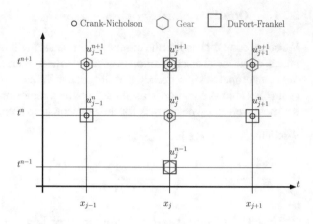

3.2.1.3 Boundary Conditions

- Dirichlet-type boundary condition corresponds to prescribed value at the boundaries

$$u(t, a) = u_a, \qquad\qquad u(t, b) = u_b,$$

where the segment $[a, b]$ is discretized using $N + 2$ nodes (thus N unknowns are determined corresponding to the interior nodes). The BTCS scheme for the first interior node reads as

$$-\mathsf{Fo}\, u_0^{n+1} + (1 + 2\mathsf{Fo})\, u_1^{n+1} - \mathsf{Fo}\, u_2^{n+1} = u_1^n \,,$$

where $\mathsf{Fo} = \frac{\alpha \Delta t}{\Delta x^2}$. As $u_0^{n+1} = u_b$, this term can be moved to the RHS (containing known terms). Performing the same procedure for the other boundary, the system for the implicit Euler scheme reads matricially as follows:

$$A\mathbf{u} = f,$$

with

$$
A = \begin{pmatrix}
1+2\mathsf{Fo} & -\mathsf{Fo} & 0 & 0 & \cdots & 0 \\
-\mathsf{Fo} & 1+2\mathsf{Fo} & -\mathsf{Fo} & 0 & \cdots & 0 \\
\vdots & & & & & \vdots \\
0 & \cdots & 0 & -\mathsf{Fo} & 1+2\mathsf{Fo} & -\mathsf{Fo} \\
0 & \cdots & 0 & 0 & -\mathsf{Fo} & 1+2\mathsf{Fo}
\end{pmatrix}
\quad
f = \begin{pmatrix} u_1^n \\ u_2^n \\ \vdots \\ u_{N-1}^n \\ u_N^n \end{pmatrix} + \begin{pmatrix} \mathsf{Fo}\, u_a \\ 0 \\ \vdots \\ 0 \\ \mathsf{Fo}\, u_b \end{pmatrix}
$$

- Neumann-type condition corresponds to prescribed normal gradient values, also called flux, at the boundaries

$$\frac{\partial u}{\partial \mathbf{n}}(a, t) = \psi_a, \qquad\qquad \frac{\partial u}{\partial \mathbf{n}}(b, t) = \psi_b,$$

where $\mathbf{n}$ represents the unit exterior normal vector at the boundary. As the BTCS scheme is of order 2 in space, a relevant approximation of the Neumann-type boundary condition should be at least of order 2 as well. Start with a space discretization of segment $[a, \ b]$ with N nodes, such that $x_1 = a$ and $x_N = b$. Consider a *ghost point* x_0 located at distance Δx before $x_1 = a$. Therefore, the Neumann boundary condition at point a can be approximated by

$$\frac{\partial u}{\partial \mathbf{n}}(a, t) \simeq -\frac{u_2^{n+1} - u_0^{n+1}}{2\Delta x}.$$

The $-$ sign is obtained from $\mathbf{n} = -1$ at point a. The matrix system associated to the Neumann boundary condition and BTCS scheme becomes

$$A\mathbf{u} = f,$$

with

$$
A = \begin{pmatrix}
1 + 2\text{Fo} & -2\text{Fo} & 0 & 0 & \ldots & 0 \\
-\text{Fo} & 1 + 2\text{Fo} & -\text{Fo} & 0 & \ldots & 0 \\
\vdots & & & & & \vdots \\
0 & \ldots & 0 & -\text{Fo} & 1 + 2\text{Fo} & -\text{Fo} \\
0 & \ldots & 0 & 0 & -2\text{Fo} & 1 + 2\text{Fo}
\end{pmatrix}
\quad
f = \begin{pmatrix}
u_1^n \\
u_2^n \\
\vdots \\
u_{N-1}^n \\
u_N^n
\end{pmatrix}
+
\begin{pmatrix}
2\text{Fo}\,\psi_a \Delta x \\
0 \\
\vdots \\
0 \\
2\text{Fo}\,\psi_b \Delta x
\end{pmatrix}
$$

- Robin-type boundary condition corresponds to mixed boundary conditions

$$\frac{\partial u}{\partial \mathbf{n}}(a, t) = \lambda_a u(a, t) + \mu_a, \qquad\qquad \frac{\partial u}{\partial \mathbf{n}}(b, t) = \lambda_b u(b, t) + \mu_b.$$

using the same technique involving a *ghost point*, the resulting matrix remains tridiagonal.

3.2.1.4 Tridiagonal Matrix Algorithm (TDMA) or Thomas Algorithm

We realized that an implicit scheme and a centered space discretization yield a matrix system. Regardless of the type of boundary condition (Dirichlet, Neumann, or Robin), a particular choice of discretization produces the matrix in its tridiagonal form.

Now, we present an efficient algorithm for solving this type of system. It is a particular case of LU decomposition that can be extended to a higher dimensional system: a tridiagonal block system. The algorithm for solving

$$Au = f$$

where

$$A = \begin{pmatrix} b_1 & c_1 & 0 & 0 & \cdots & 0 \\ a_2 & b_2 & c_2 & 0 & \cdots & 0 \\ \vdots & & & & & \vdots \\ 0 & \cdots & 0 & a_{N-1} & b_{N-1} & c_{N-1} \\ 0 & \cdots & 0 & 0 & a_N & b_N \end{pmatrix} \quad \text{and} \quad f = \begin{pmatrix} f_1 \\ f_2 \\ \vdots \\ f_{N-1} \\ f_N \end{pmatrix}$$

corresponds to the following steps:

1. Divide the first row of the matrix system by b_1 such that the first equation reads as

$$u_1 + \gamma_1 u_2 = \rho_1 ,$$

where $\gamma_1 = c_1/b_1$ and $\rho_1 = f_1/b_1$.
2. Subtract the new first row times a_2 to the second row of the matrix system. Then divide by $b_2 - a_2\gamma_1$ such that the second row becomes

$$u_2 + \gamma_2 u_3 = \rho_2,$$

where $\gamma_2 = \dfrac{c_2}{b_2 - a_2\gamma_1}$ and $\rho_2 = \dfrac{f_2 - a_2\rho_1}{b_2 - a_2\gamma_1}$.
3. The repetition of this process until the last row yields

$$\begin{pmatrix} 1 & \gamma_1 & 0 & 0 \cdots & 0 \\ 0 & 1 & \gamma_2 & 0 \cdots & 0 \\ \vdots & & & & \vdots \\ 0 \cdots & 0 & 0 & 1 & \gamma_{N-1} \\ 0 \cdots & 0 & 0 & 0 & 1 \end{pmatrix} \begin{pmatrix} u_1 \\ u_2 \\ \vdots \\ u_{N-1} \\ u_N \end{pmatrix} = \begin{pmatrix} \rho_1 \\ \rho_2 \\ \vdots \\ \rho_{N-1} \\ \rho_N \end{pmatrix}.$$

Here, for $i = 2, \ldots, N - 1$,

$$\gamma_i = \frac{c_i}{b_i - a_i\gamma_{i-1}}, \qquad \rho_i = \frac{f_i - a_i\rho_{i-1}}{b_i - a_i\gamma_{i-1}},$$

and

$$\gamma_1 = c_1/b_1, \qquad \rho_1 = f_1/b_1, \qquad \rho_N = \frac{f_N - a_N\rho_{N-1}}{b_N - a_N\gamma_{N-1}}.$$

4. Thus, the computation of **u** is straightforward from the bottom to top as

$$u_N = \rho_N$$
$$u_i = \rho_i - \gamma_i u_{i+1}, \qquad\qquad \text{for } i = N-1, \dots, 1.$$

3.2.2 Nonlinear Case

Now, we focus on a diffusion coefficient depending on the variable:

$$\frac{\partial u}{\partial t} = \frac{\partial}{\partial x}\left(\alpha(u)\frac{\partial u}{\partial x}\right), \qquad (3.13)$$

where we assume that α remains positive and that for a given u, there is only one associated value for α.

Strategies for solving the *nonlinearity* often involve calling a root function.

3.2.2.1 Kirchoff's Transformation

The first idea is to devise the diffusion coefficient from the Laplacian operator such that Eq. (3.13) becomes an equation associated to a linear differential operator. This can be accomplished using the Kirchoff change of variable:

$$v \equiv \mathcal{K}(u) = \int_{u^\star}^{u} \alpha(w)\,dw,$$

where $u^\star$ is any reference value of u. Thus, Eq. (3.13) is transformed in

$$\frac{1}{[\alpha(u) - \alpha(u^\star)]^{-1}}\frac{\partial v}{\partial t} = \frac{\partial^2 v}{\partial x^2}.$$

The new diffusion term $[\alpha(u) - \alpha(u^\star)]$ must be computed in function $u = \mathcal{K}^{-1}(v)$, and then Kirchoff's formula must be explicated. Moreover, the boundary and initial conditions must be transformed as

- Dirichlet: $v_D = \mathcal{K}(u_D)$,
- Neumann: $\dfrac{\partial v}{\partial n} = \alpha(u)\dfrac{\partial u}{\partial n}$,
- Robin: $\dfrac{\partial v}{\partial n} = -h\left(\mathcal{K}^{-1}(v) - u\right)$.

For example, for a linear variation of conductivity

$$\alpha = \alpha_0\left(1 + \beta u\right),$$

the Kirchoff transformation (with $u^* = 0$) reads as

$$\mathscr{K}(u) = v = \alpha_0 \left(u + \frac{\beta}{2} u^2 \right),$$

and its inverse is simply

$$u = \mathscr{K}^{-1}(v) = \frac{1}{\beta} \left(\sqrt{1 + 2\beta v} - 1 \right).$$

From here, any standard numerical scheme can be used.

1. The diffusion coefficient is computed as $\alpha(u) = \alpha_0 (1 + \beta u)$. As the gauge can be arbitrarily taken, we choose $u^* = 0$; therefore, $[\alpha(u) - \alpha(u^*)] = \alpha_0 \beta u$.
2. Choose any method for computing

$$\frac{\partial v}{\partial t} = (\alpha_0 \beta u) \frac{\partial^2 v}{\partial x^2}.$$

3. Compute back $u = \dfrac{1}{\beta} \left(\sqrt{1 + 2\beta v} - 1 \right)$ and proceed as such.

Remark 8 This transformation can easily be used for multidimensional space cases.

3.2.2.2   Coupled System Approach

Let us consider Eq. (3.13) and introduce the flux density $q = \alpha(u) \frac{\partial u}{\partial x}$ so that the heat equation system becomes

$$\frac{\partial u}{\partial t} = \frac{\partial q}{\partial x} \qquad \text{and} \qquad q = \alpha(u) \frac{\partial u}{\partial x}. \tag{3.14}$$

A natural discretization for Eq. (3.14) by using the general θ scheme reads

$$\frac{u_j^{n+1} - u_j^n}{\Delta t} = \theta \left[\frac{q_{j+1/2}^{n+1} - q_{j-1/2}^{n+1}}{\Delta x} \right] + (1 - \theta) \left[\frac{q_{j+1/2}^n - q_{j-1/2}^n}{\Delta x} \right]$$

and

$$q_{j\pm 1/2} = \pm \alpha(u_{\pm 1/2}) \frac{u_{j\pm 1} - u_j}{\Delta x}.$$

The choice of the time level for computing the two variables (u, q) is important mainly in the definition of the flux term and the spatial discretization involving half-node points for the flux. Thus, two alternatives are considered: choose a second grid of computation, one for each dependent variable, or approximate the mid-point field by the original nodes. Another disadvantage of this discretization is to *artificially* define the boundary conditions for the flux because they are not defined at the continuous level (thus, this numerical modeling has an inherent error due to the approximations of the boundary conditions for the flux density q).

3.2.2.3 Iterative Method

In practice, iterative methods are most widely used for solving nonlinear terms wherever they occur (through a source term, through thermal dependency properties, in the boundary conditions). By solving a nonlinear equation

$$\mathscr{L}(u) = 0,$$

using a fixed-point strategy involves

1. making a *first guess* for $u^{(k=0)}$, and computing the next evaluation of the unknown $u^{(k+1)}$ for the linear equation

$$u^{(k+1)} = \mathscr{B}u^{(k)} + C,$$

where $\mathscr{B}$ is closed to the linearization of $\mathscr{L}$. The tools devoted are often called *root-finding algorithms*, because the following function is considered: $g(x) \equiv x - [\mathscr{B}x + C]$. The main strategies for obtaining this are

- Bracketing methods: here, the convergence of the methods is guaranteed (but to which root is another dilemma). The rate of convergence of the algorithm is very slow. The most popular is the *dichotomy or bisection* algorithm.
- Newton-like algorithms: these are the most popular and are based on the iterative formulation

$$x^{(k+1)} = x^{(k)} - \frac{g(x^{(k)})}{g'(x^{(k)})}.$$

The estimation of the derivative g' yields different Newton-like methods.

2. Choosing a tolerance η and a stopping criteria, for which the root-finding algorithm is stopped. Most of the time, the stopping criteria is chosen to be

$$\frac{\left| u^{(k+1)} - u^{(k)} \right|}{\left| u^{(k)} \right|} \leqslant \eta.$$

3.2.3 Applications in Engineering

The following sections present some applications.

3.2.3.1   Room Air Temperature Simulation

Mathematical Modeling

Consider a single room, with time-dependent temperature T given by Eq. (3.1) (Chap. 2, Sect. 2.9):

$$\rho_a c_a V_a \frac{\mathrm{d}T}{\mathrm{d}t} = -BT + A.$$

We already know many efficient schemes for numerically solving this model. Recall that term A includes the surrounding effects. In particular, the spatial boundary conditions, corresponding to the walls, roof, and ground temperatures, are contained in this term. Thus, roughly speaking, coefficient A should be a function of time $A(t)$, or more precisely of the surrounding temperatures, noted generically as $\tilde{A}(\mathbf{u}) \overset{\text{def}}{:=} A(t)$ because it evolves. The model for indoor temperature becomes

$$\begin{aligned}
\rho_a c_a V_a \frac{\mathrm{d}T}{\mathrm{d}t} &= -BT + A(t) \\
&= -BT + \tilde{A}(\mathbf{u}).
\end{aligned}$$

To complete the modeling, the thermal energy passing through each wall, roof, and ground must also be computed, assuming the exterior conditions are known. Thus, for each surface, a heat equation must be solved. Let us consider the simplified modeling for which the temperature field travels only in a transversal direction, reducing the 3D geometry to a 1D configuration. Thus, one must deal with six (one for each face) times the heat equation in 1D:

$$\mathscr{H}_i(\mathbf{u}) = \Psi_i(T, T^\infty),$$

where $\mathscr{H} \equiv \partial_t - \alpha \partial_{xx}^2$ is the heat operator and Ψ is the boundary operator involving both the exterior temperature T^∞ and the indoor room temperature T. The heat field within all the walls is noted as $\bar{u}$, as before. As in previous sections, we presented many relevant numerical schemes for simulating these equations; therefore, we do not need to repeat it again but rather emphasize about the algorithm strategy.

Numerical Algorithm

There are three main strategies to solve the indoor room air temperature

$$\rho_a c_a V_a \frac{\mathrm{d}T}{\mathrm{d}t} = -BT + \tilde{A}(\mathbf{u}),$$

and the surrounding thermal flux crossing the building

$$\mathscr{H}_i(\mathbf{u}) = \Psi_i(T, T^\infty): \qquad\qquad \text{for}\quad i = 1, \dots, 6.$$

- First, compute the thermal flux through the walls, then compute the indoor air room. Therefore, the semidiscretization reads as

$$
\begin{cases}
\mathscr{H}_i(\mathbf{u}^{n+1}) = \Psi_i(T^n, T^\infty) & \text{for } i = 1, \ldots, 6, \\
\rho_a c_a V_a \dfrac{T^{n+1} - T^n}{\Delta t} = -B\, T^{n+1} + \tilde{A}(\mathbf{u}^{n+1}).
\end{cases}
$$

Note that the thermal diffusion through the walls must be explicit because the internal temperature occurring in the boundary condition is given at time t^n. In contrast, the discrete equation for T is computed using an implicit method. The contrary would yield a physically inconsistent model. This strategy is limited by the stability condition imposing a time-step threshold associated to the PDEs.

- First, compute the indoor air room, then compute the thermal flux through the walls. The corresponding semidiscretization reads as

$$
\begin{cases}
\rho_a c_a V_a \dfrac{T^{n+1} - T^n}{\Delta t} = -B\, T^n + \tilde{A}(\mathbf{u}^n), \\
\mathscr{H}_i(\mathbf{u}^{n+1}) = \Psi_i(T^{n+1}, T^\infty), & \text{for } i = 1, \ldots, 6
\end{cases}
$$

Here, the first discrete equation must be explicit, else the model inherits some physical inconsistency. In addition, the thermal diffusion through the walls must be implicit for similar reasons. Finally, this strategy is limited by the stability condition imposing a time-step threshold associated to the ODE.

- The other strategy is to compute everything by using only implicit schemes:

$$
\begin{cases}
\rho_a c_a V_a \dfrac{T^{n+1} - T^n}{\Delta t} = -B\, T^{n+1} + \tilde{A}(\mathbf{u}^{n+1}) \\
\mathscr{H}_i(\mathbf{u}^{n+1}) = \Psi_i(T^{n+1}, T^\infty), & \text{for } i = 1, \ldots, 6.
\end{cases}
$$

Thus, it is necessary to call some iterative routines to ensure numerical consistency of the system of discrete equations. As shown earlier, the 1D heat equation can be solved with an *optimal*[10] numerical cost by using the Thomas algorithm. Moreover, the first guess and the fixed-point method may be accurately chosen such that the subiterations within a time step for holding the system completely implicit may not be too expansive. Thus, this strategy gains robustness and may not loose considerably in computation cost.

- The last strategy involves the computing of all the fields with explicit schemes.

$$
\begin{cases}
\rho_a c_a V_a \dfrac{T^{n+1} - T^n}{\Delta t} = -B\, T^n + \tilde{A}(\mathbf{u}^n), \\
\mathscr{H}_i(\mathbf{u}^{n+1}) = \Psi_i(T^n, T^\infty) & \text{for } i = 1, \ldots, 6.
\end{cases}
$$

---

[10]The number of operations is comparable as for an explicit system of equations.

Obviously, this strategy is the most time consuming because of the severe limitation of the admissible time-step.

Remark 9 All the four algorithm presented earlier are distinctive: there is no guarantee that the numerical solutions coincide. Moreover, it would be unusual for them to produce the same results.

Numerical Algorithm: MTDMA

The algorithm for efficiently solving the MTDMA system corresponds to a natural extension of the TDMA algorithm because it is in fact a Gauss pivoting algorithm. Rename the block matrix such that, for all well-defined j,

$$A_j = M_{j-1,j}, \qquad B_j = M_{j,j}, \qquad C_j = M_{j,j+1}.$$

Thus, we have a formal tridiagonal block-matrix system, where the subdiagonal is $\{A_j\}$, the diagonal is $\{B_j\}$, and the upper diagonal is $\{C_j\}$. Hereafter, the superscript $n+1$ of u is dropped.

1. Assume there exists for $j = 0$ to $N - 1$ relations such that

$$u_j = P_j\, u_{j+1} + q_j.$$

In particular, for $j = 0$, one has

$$u_0 = B_0^{-1}\left[-C_0 u_1 + f_0\right].$$

Thus,

$$P_0 = -B_0^{-1} C_0, \qquad \text{and} \qquad q_0 = B_0^{-1} f_0.$$

2. Further, for $j = 1$ to $N - 1$,

$$A_j\, u_{j-1} + C_j\, u_j + C_j\, u_{j+1} = f_j.$$

By injecting the relation $u_{j-1} = P_{j-1}\, u_j + q_{j-1}$ for $j = 1, \ldots, N$, we have

$$A_j\left(P_{j-1} u_j + q_{j-1}\right) + C_j\, u_j + C_j\, u_{j+1} = f_j, \tag{3.15}$$

that can be recast in the following, for $j = 1, \ldots, N$:

$$u_j = \left[A_j\, P_{j-1}^{-1} + B_j\right]^{-1}\left(-C_j u_{j+1} + (f_j - A_j\, q_{j-1})\right).$$

Thus, by identifying terms by using Eq. (3.15), we obtain the two recurrence sequences that must be verified for $j = 1, \ldots, N - 1$:

$$P_j = -\left[A_j \, P_{j-1}^{-1} + B_j\right]^{-1} C_j,$$

$$q_j = \left[A_j \, P_{j-1}^{-1} + B_j\right]^{-1} \left(f_j - A_j \, q_{j-1}\right).$$

As P_0 and q_0 are known explicitly, all the terms can be recursively computed until P_{N-1} and q_{N-1}.

3. Next, we may use the known matrix $\{P_j\}$ and vectors $\{q_j\}$ to compute each u_j. For this, as for the Thomas algorithm, the computation begins from the last node $j = N$ and continues back until the first node by using the relation

$$u_{j-1} = P_{j-1} u_j + q_{j-1},$$

for $j = 1, \ldots, N$. Specifically, for the last nodes, $j = N$, the abovementioned relation is

$$u_{N-1} = P_{N-1} u_N + q_{N-1},$$

and by comparing with the last equations of the matrix system

$$u_{N-1} = A_N^{-1} \left[-B_N u_N + f_N\right],$$

one can explicitly compute both u_N and u_{N-1} simultaneously. This allows to initiate the backward computation of all unknowns $\{u_j\}$.

3.2.4 Heat Equation in Two and Three Space Dimensions

Consider once again the heat equation in higher space dimension. In the following, the description is made for 2D case but the extension for 3D is natural. However, the numerical properties are precise for both cases. Thus, consider

$$\frac{\partial u}{\partial t} = \alpha \left(\frac{\partial^2 u}{\partial x^2} + \frac{\partial^2 u}{\partial y^2}\right), \tag{3.16}$$

with some boundary conditions $\mathscr{B}(u) = 0$.

3.2.4.1 Standard Discretization

Here, we simply preform the computation of the 2D heat equation by using a Crank–Nicolson scheme. This will be used as a reference but it is known to be very time consuming even if its accuracy remains of order $\mathcal{O}(\Delta t^2 + \Delta x^2 + \Delta y^2)$. Introduce the following finite difference operator for any variable v_{ijk}:

$$D_x^2 v_{ijk} = \frac{1}{\Delta x^2} \left(v_{i-1,jk} - 2v_{ijk} + v_{i+1,jk} \right),$$

$$D_y^2 v_{ijk} = \frac{1}{\Delta y^2} \left(v_{i,j-1,k} - 2v_{ijk} + v_{i,j+1,k} \right),$$

$$D_z^2 v_{ijk} = \frac{1}{\Delta z^2} \left(v_{ij,k-1} - 2v_{ijk} + v_{ij,k+1} \right).$$

The Crank–Nicolson discretization in 3D reads as

$$u_{ijk}^{n+1} = u_{ijk}^n + \alpha \frac{\Delta t}{2} \left[D_x^2 + D_y^2 + D_z^2 \right] \left(u_{ijk}^{n+1} + u_{ijk}^n \right).$$

At each node (ijk), this scheme involves seven unknowns in 3D and five unknowns in 2D. Thus, the corresponding matrices are respectively 7-banded and 5-banded sparse matrices. Of course, the Thomas algorithm cannot be applied directly. Even if the inverse matrix algorithm is chosen carefully, the computational cost may be prohibitive for large-scale problems.

3.2.4.2   Locally 1D Method

The idea of the Locally 1D (LOD) method is to split the time step into stages, and compute for each stage the diffusion process in only one spatial direction. Numerically, this allows to compute only tridiagonal matrix systems (that can be realized optimally by using the Thomas algorithm). The locally 1D method reads as follows:

1. Compute the intermediate stage value $\bar{u}_{ij}$ through a diffusion in the x-direction:

$$\bar{u}_{ij} = u_{ij}^n + \alpha \frac{\Delta t}{2} \left(D_x^2 \bar{u}_{ij} + D_x^2 u_{ij}^n \right).$$

2. Finally, compute u_{ij}^{n+1} through a diffusion in the y-direction:

$$u_{ij}^{n+1} = \bar{u}_{ij} + \alpha \frac{\Delta t}{2} \left(D_y^2 u_{ij}^{n+1} + D_y^2 \bar{u}_{ij} \right).$$

An important issue focuses on the boundary conditions used for the computation of the intermediate value $\bar{u}_{ij}$. Even if the time step used for each stage is exactly half the entire time step Δt, it is not true that $\bar{u}_{ij}$ corresponds to the heat field at time $\frac{\Delta t}{2}$. Thus, it is not relevant (and even consistent) to take $\mathcal{B}(\bar{u}) = \mathcal{B}(u^{n+1/2})$ as boundary conditions.

However, this method is very efficient when using *well-behaved* boundary conditions as null flux or constant in time Dirichlet boundary conditions. Owing to its robustness (unconditional stability), the method's second-order accuracy[11] is *easy to implement*, and the method shows an optimal time computational cost, thus being an interesting numerical scheme. Moreover, the extension to the 3D case is straightforward.

3.2.4.3 Alternating Direction Implicit Method

Here, once again, the idea is to split the time step into intermediate stages. However, contrary to the LOD method, in the Alternating Direction Implicit (ADI) method, diffusion in all directions is computed.

Peaceman–Rachford   This scheme proposed by Peaceman & Rachford is as follows

1. Compute the intermediate stage value $\bar{u}_{ij}$ associated to a diffusion in x-direction, including a y-direction diffusion for the former value u^n as

$$\bar{u}_{ij} = u_{ij}^n + \alpha \frac{\Delta t}{2} \left(D_x^2 \bar{u}_{ij} + D_y^2 u_{ij}^n \right).$$

2. Next, compute u_{ij}^{n+1} through a diffusion in the y-direction, considering an x-direction diffusion for the intermediate value $\bar{u}_{ij}$:

$$u_{ij}^{n+1} = \bar{u}_{ij} + \alpha \frac{\Delta t}{2} \left(D_y^2 u_{ij}^{n+1} + D_x^2 \bar{u}_{ij} \right).$$

Here, once again we must solve the TDMA system; this is done very efficiently by using the Thomas algorithm. In addition, the accuracy is theoretically of order $\mathscr{O}(\Delta t^2 + \Delta x^2 + \Delta y^2)$. Moreover, the intermediate value $\bar{u}$ effectively corresponds to the heat field at the half time $\frac{\Delta t}{2}$; therefore, defining the boundary conditions should not be an issue (for example, we can extrapolate the boundary conditions from previous times). However, the Peaceman & Rachford ADI scheme does not conserve its numerical properties when extended to a 3D case; instead, it becomes *only* first-order accurate $\mathscr{O}(\Delta t + \Delta x^2 + \Delta y^2 + \Delta z^2)$, and turns conditionally stable.

Douglas–Gunn   In 3D, Douglas & Gunn proposed a second-order accurate scheme $\mathscr{O}(\Delta t^2 + \Delta x^2 + \Delta y^2)$ that also remains unconditionally stable.

1. Compute a first stage value $\bar{u}$ by using

   - the diffusion in x direction for both $\bar{u}$ and u^n and
   - the (x, y, z)-direction diffusion computing with the current value of heat field u^n

---

[11]By alternating the order of the direction computed at first, insuring the *commutativity of the operators*, the theoretical accuracy can increase by one order for the time step.

This yields the following first stage:

$$\bar{u}_{ijk} = u^n_{ijk} + \alpha \frac{\Delta t}{2} D^2_x \left(\bar{u}_{ijk} + u^n_{ijk} \right) + \alpha \Delta t D^2_y \left(u^n_{ijk} \right) + \alpha \Delta t D^2_z \left(u^n_{ijk} \right).$$

2. Compute a second stage value $\hat{u}$ through

   - the diffusion in y direction involving $\hat{u}$ and u^n,
   - the diffusion in x direction for both $\bar{u}$ and u^n, with $\bar{u}$ given known, and
   - the z-direction diffusion computing u^n, as for the first stage.

   That is,

$$\hat{u}_{ijk} = u^n_{ijk} + \alpha \frac{\Delta t}{2} D^2_x \left(\bar{u}_{ijk} + u^n_{ijk} \right) + \alpha \frac{\Delta t}{2} D^2_y \left(\hat{u}_{ijk} + u^n_{ijk} \right) + \alpha \Delta t D^2_z \left(u^n_{ijk} \right).$$

3. Compute the final stage for u^{n+1} through

   - the z-direction diffusion involving u^{n+1} and u^n,
   - the diffusion in x-direction for both the known variables $\bar{u}$ and u^n, and
   - the diffusion in y-direction, with known variables $\hat{u}$ and u^n.

   Therefore, the third stage reads as

$$u^{n+1}_{ijk} = u^n_{ijk} + \alpha \frac{\Delta t}{2} D^2_x \left(\bar{u}_{ijk} + u^n_{ijk} \right) + \alpha \frac{\Delta t}{2} D^2_y \left(\hat{u}_{ijk} + u^n_{ijk} \right) + \alpha \frac{\Delta t}{2} D^2_z \left(u^n_{ijk} + u^{n+1}_{ijk} \right).$$

As the intermediate values of the unknown values $\bar{u}$ and $\hat{u}$ correspond to the evaluation of the heat field at times $t^n + \dfrac{\Delta t}{3}$ and $t^n + \dfrac{2\Delta t}{3}$ respectively, it is easy to associate relevant boundary conditions for these intermediate values.

Other schemes: There are numerous 3D unconditionally stable methods with second-order accurate schemes. For example, Dyakonov, Yanenko, and the fractional step methods Douglas–Gunn for multistep procedures.

3.2.5 Exercises

Theoretical

Crank–Nicolson Scheme for Convective Boundary Conditions

Write down the tridiagonal matrix system corresponding to the Crank–Nicolson scheme for 1D heat transfer by using Robin boundary conditions.

Multilayer Medium

Consider the multilayer medium of thermal conductivities λ_i. Recall the linear heat equation:

$$\frac{\partial}{\partial t} u(x, t) = v_i \frac{\partial^2}{\partial x^2} u(x, t), \tag{3.17}$$

where u is the heat field, depending on time t and space x. The diffusion coefficient $v_i = \dfrac{\lambda_i}{\rho C}$ depends on the layer. Specifically, assume a space discretization such that there is a node x_i located exactly at the interface of two layers A and B of thermal conductivities λ_A and λ_B, respectively. What is the equation at that point such that the continuity of both heat flux density and heat field are preserved?

1. By using the Taylor expansion, show that

$$\left(\frac{\partial^2 u}{\partial x^2}\right)_{iA} \simeq \frac{2}{(\Delta x)^2}\left[u_{i-1} - u_i + \Delta x \left(\frac{\partial u}{\partial x}\right)_{iA}\right].$$

2. Show that Eq. (3.17) for the A-layer reads

$$\Delta x \left(\frac{\partial u}{\partial x}\right)_{iA} = \frac{(\Delta x)^2}{2v_A} \frac{u_i^{n+1} - u_i^n}{\Delta t} + (u_i - u_{i-1}).$$

3. Analogously, for the B-layer, show that

$$-\Delta x \left(\frac{\partial u}{\partial x}\right)_{iB} = \frac{(\Delta x)^2}{2v_B} \frac{u_i^{n+1} - u_i^n}{\Delta t} + (u_i - u_{i+1}).$$

4. Recall that the continuity of the heat flux density at the interface is

$$\lambda_A \left(\frac{\partial u}{\partial x}\right)_{iA} = \lambda_B \left(\frac{\partial u}{\partial x}\right)_{iB}.$$

   Deduce the time evolution of the temperature at the interface as follows:

$$u_i^{n+1} = u_i^n + \left[\frac{2v_A \Delta t/(\Delta x)^2}{v_A/v_B + \lambda_A/\lambda_B}\right]\left[u_{i+1} - \left(1 + \frac{\lambda_A}{\lambda_B}\right)u_i + \frac{\lambda_A}{\lambda_B}u_{i-1}\right].$$

5. Verify that when the thermal conductivities are the same ($\lambda_A = \lambda_B$), the aforementioned relation coincides with a standard discretization of the heat equation for a homogeneous medium.

Comparison with Finite Volume Method in 1D

Consider the 1D steady-state diffusion equation

$$-u''(x) = f(x) \qquad\qquad \text{for } x \in]0, 1[, \tag{3.18}$$

with Dirichlet boundary conditions $u(0) = u(1) = 0$.

1. Draw a diagram with segment $[0, 1]$ divided in N cells K_i, $i = 1, \ldots, N$ of size h_i (a priori all different). The centers of cells K_i are denoted by x_i. The distance between two neighboring cells x_i and $x_{i\pm1}$ is denoted as $h_{i\pm1/2}$. The points located at the boundary with two neighboring cells K_i and $K_{i\pm1}$ are denoted by $x_{i\pm1/2}$.

2. Integrate Eq. (3.18) on each cell K_i and show that N equations are obtained that read as

$$F_{i+1/2} - F_{i-1/2} = h_i f_i \qquad\qquad \text{for } x \in]0, 1[$$

where $f_i = \frac{1}{h_i} \int_{K_i} f(x) \, \mathrm{d}x$ and $F_{i+1/2} = -u'(x_{i+1/2})$. The term $F_{i+1/2}$ is called the flux at the interface $x_{i+1/2}$.

3. By approximating the $N + 1$ fluxes at the interfaces by using N unknowns u_i and using the boundary conditions, show that the numerical approximations read as

$$F_{i+1/2} \simeq -\frac{u_{i+1} - u_i}{h_{i+1/2}} .$$

4. How to discretize the flux at $x_{i-1/2}$?

5. Show that the system obtained by the discretization of Eq. (3.18) by using the finite volume method can be cast into the matrix form as

$$A\mathbf{u} = b,$$

where

$$(A\mathbf{u})_i = \frac{1}{h_i} \left(-\frac{u_{i+1} - u_i}{h_{i+1/2}} + \frac{u_i - u_{i-1}}{h_{i-1/2}} \right), \qquad\qquad b_i = f_i,$$

6. Compare the approximation of $u''(x_i)$ through a finite difference approach by using a non-regular space discretization.

7. We want to show that the approximation of $u''(x_i)$ by using

$$\frac{1}{h_i} \left(\frac{u(x_{i+1}) - u(x_i)}{h_{i+1/2}} - \frac{u(x_i) - u(x_{i-1})}{h_{i-1/2}} \right)$$

is not *consistent*, considering the finite difference approach, that is, R_i defined as

$$R_i = \frac{1}{h_i} \left(\frac{u(x_{i+1}) - u(x_i)}{h_{i+1/2}} - \frac{u(x_i) - u(x_{i-1})}{h_{i-1/2}} \right) - u''(x_i),$$

does not tend, in general, toward 0 when h_i tends toward 0.

a. By using the Taylor expansion, show that

$$R_i = \left(\frac{1}{h_i} \frac{h_{i+1/2} + h_{i-1/2}}{2} - 1 \right) u''(x_i) + O\left(\frac{h_{i-1/2}^2 - h_{i+1/2}^2}{h_i} \right).$$

b. What is the condition for $R_i \to 0$ when $h_i \to 0$?

c. What is the R_i tendency, for example, when the space steps are taken as $h_i = h$ for even i, and $h_i = h/2$ for odd i?

Chapter 4
Basics in Practical Finite-Element Method

This chapter is devoted to a practical presentation of the finite-element method (FEM). The focus is on the construction of numerical schemes rather than on the numerical properties that this approach benefits; References [5, 106] provide an introduction. A very large literature survey, sorted by fundamental references, mathematical foundations, applications, implementation techniques, and other special topics as well as proceedings of symposia and conferences, can be found in [146].

4.1 Heat Equation

We start with the introduction of the simple steady-state one-dimensional heat equation, so that all the steps for constructing a finite element scheme can be easily computed. The main steps can thus be highlighted, and the presentation for a general case would be smoother. Moreover, considering finite-difference discretization, a straightforward semi-discretization on the unsteady heat equation yields to the Poisson equation plus a source term depending on the variable; this is also known as the Helmholtz equation. Thus,

$$-\frac{d^2 u}{dx^2}(x) = f(x) \qquad \text{for} \quad x \in [a, b], \qquad (4.1)$$

with the Dirichlet boundary conditions $u(a) = u_a$ and $u(b) = u_b$.

Remark 4.1 The minus sign $(-)$ in front of the differential operator is used as a convenient way to obtain a weak formulation with a plus sign $(+)$, as we will see hereafter.

The steps for constructing a finite element scheme are relatively different from those of the finite difference discretization. First, the solution u of Eq. (4.1), with boundary conditions, is said to be a *strong solution* related to the *strong formulation* of the Problem (4.1): here we seek a solution that should be twice differentiable.

© Springer Nature Switzerland AG 2019
N. Mendes et al., *Numerical Methods for Diffusion Phenomena in Building Physics*,
https://doi.org/10.1007/978-3-030-31574-0_4

Instead of searching for such a strong solution, we associate a *weak formulation* (i.e. a variational formulation) to the Poisson problem (cf. Sect. 4.1.1). Thus, we require that the solution should contain a *weaker* regularity.[1] The weak solution is then approximated by assuming a decomposition on a well-chosen functional basis

$$u(x) = \sum \bar{u}_i \phi_i(x).$$

This basis of function ϕ_{ii}, added to a partition of the domain by 1D segments, is called a set of *finite elements* (cf. Sect. 4.1.2). By injecting this functional approximation into the weak formulation, we obtain a system of equations whose unknowns are the weights $\mathbf{u}_{ii}$ occurring in the functional decomposition of the weak solution (cf. Sect. 4.1.3).

4.1.1 Weak Formulation and Test Functions

The variational formulation of Eq. (4.1) is obtained by *projecting* the operator onto some test functions ϕ defined in $[a, b]$:

$$\int_a^b \left[-\frac{d^2 u}{dx^2}(x) \right] \cdot \phi(x)\, dx = \int_a^b [f(x)] \cdot \phi(x)\, dx.$$

Remark 4.2 The projection of a given operator $\mathcal{L}$ onto ϕ over domain Ω involves the multiplication of the operator by ϕ, and then its integration over Ω. This is understood as the generalization of the scalar product for vectors (on a finite dimensional space) on a space of functions with infinite dimensions.

The integration by parts easily provides

$$-\left[\frac{du}{dx}(x) \cdot \phi(x) \right]_a^b + \int_a^b \left(\frac{du}{dx}(x) \cdot \frac{d\phi}{dx}(x) \right) dx = \int_a^b (f(x) \cdot \phi(x))\, dx.$$

Note that the order of the differential operator for u has decreased by one. Thus, integrating by parts allows us to search for a solution u derivable only once (instead of twice). Moreover, the integral terms on both sides must have finite values. Specifically, the first derivatives u' and ϕ' must be *integrable*[2] on segment $[a, b]$. The space of such functions is given as $\mathcal{H}^1([a, b])$.

In addition, to remove the boundary term, as we do not have any information on $u'(x)$ evaluated at the boundaries a and b (because of the Dirichlet-type boundary

---

[1] A mathematical demonstration is needed to prove that the *weak* solution, associated to the variational formulation of the problem, is also a solution of the original *strong* formulation of the Poisson problem.

[2] The notion of a derivative must be understood in a weak sense, and the notion of integrability must be considered in the L^2 sense.

conditions), we may consider the test function ϕ to have a zero-value at points $\phi(a) \equiv \phi(b) \equiv 0$. Thus, by using $\mathcal{H}_0^1([a, b])$, let us consider the subset of $\mathcal{H}^1([a, b])$ of functions with a null trace[3] on the boundary points a and b. Finally, the weak formulation of the problem associated to Eq. (4.1) with Dirichlet boundary conditions can be stated as follows:

Find $u \in \mathcal{H}^1([a, b])$ as the solution of

$$\int_a^b \left(\frac{du}{dx}(x) \cdot \frac{d\phi}{dx}(x) \right) dx = \int_a^b (f(x) \cdot \phi(x)) \, dx \quad \forall \phi \in \mathcal{H}_0^1([a, b]), \quad (4.2)$$

such that $u(a) = u_a$ and $u(b) = u_b$.

4.1.2 Finite Element Representation

Here we show the main distinction with the finite differences method: instead of approximating the operators of the problem (the variational formulation), we retain the operators, and instead look for approximations of the solution space. Recall that u and ϕ belong to some functional space ($\mathcal{H}^1$ and $\mathcal{H}_0^1$) with infinite dimensions (i.e. the number of functions that compose the basis is infinite). Thus, we define an approximation space $\mathcal{V}_h$ of finite dimensions (i.e. there is a finite number of elements that generate other elements of the same space) and determine an approximation u_h of the weak solution u, considering the variational formulation applied on $\mathcal{V}_h$:

- Consider a partition of $[a, b]$ in N_K segments $\{K_i\}_{i=1,...,N_K}$. For fixing ideas and simple explicit computation, assume $N_K = 5$ such that the segments are as follows:

$$K_1 = [x_0 = a, \ x_1], \qquad K_2 = [x_1, \ x_2], \qquad K_3 = [x_2, \ x_3],$$
$$K_4 = [x_3, \ x_4], \qquad K_5 = [x_4, \ x_5 = b]$$

  All the segments have the same length $| K_i | = \Delta x$.
- In addition, let us define a set of points $\{x_j\}_{j=1,...,N_x}$ on the entire segment $[a, b]$ to correspond to the locations of the discrete unknowns (i.e. the degree of freedom of the system). In our example, we selected six nodes $\{x_0; x_1; x_2; x_3; x_4; x_5\}$ that coincide with the extremities of each segment K_i. Other similar sets of degrees-of-freedom include the centers of each K_i.
- Next, define an approximation space $\mathcal{V}_h$ for $\mathcal{H}^1([a, b])$ (the space where the weak solution u is pursued; Eq. (4.2)) using the sets defined earlier. A simple choice for $\mathcal{V}_h$ exists in the set of piecewise-linear functions, with a basis composed by the piecewise linear functions ϕ_i^h that assume the value one on node x_i, and zero otherwise. In the partition $[a, b]$ defined by K_i, where $i = 1, \ldots, N_K = 5$ segments, and for each of the nodes $x_0, \ldots, x_5$, consider the corresponding functions $\phi_0^h(x), \ldots, \phi_5^h(x)$ such that $\phi_i^h(x_j) = \delta_{ij}$, for $i, j = 0, \ldots, 4$, where $\delta_{ij} = 1$ if

---

[3]It is difficult to define the notion of null trace. For our purposes, let us consider its intuitive meaning.

$i = j$, otherwise $\delta_{ij} = 0$. Moreover, a linear interpolation is assumed between $\phi_i^h(x_i)$ and $\phi_i^h(x_{i\pm1})$. This *hat functions basis* is conventionally called a $\mathbb{P}_1$ function.

Remark 4.3 It is not obvious that the u_h obtained is an approximation of u. The sense of *approximation* has to be defined clearly. Moreover, the choice of $\mathcal{V}_h$ is not obvious.

Explicitly, the basis functions are as shown in Fig. 4.1:

$$\phi_0^h(x) = \begin{cases} \dfrac{x_1 - x}{\Delta x} & \text{for } x \in [x_0, x_1] \\ 0 & \text{otherwise} \end{cases} \qquad \phi_1^h(x) = \begin{cases} \dfrac{x - x_0}{\Delta x} & \text{for } x \in [x_0, x_1] \\ \dfrac{x_2 - x}{\Delta x} & \text{for } x \in]x_1, x_2] \\ 0 & \text{otherwise} \end{cases}$$

and

$$\phi_2^h(x) = \begin{cases} \dfrac{x - x_1}{\Delta x} & \text{for } x \in [x_1, x_2] \\ \dfrac{x_3 - x}{\Delta x} & \text{for } x \in]x_2, x_3] \\ 0 & \text{otherwise} \end{cases} \qquad \phi_3^h(x) = \begin{cases} \dfrac{x - x_2}{\Delta x} & \text{for } x \in [x_2, x_3] \\ \dfrac{x_4 - x}{\Delta x} & \text{for } x \in]x_3, x_4] \\ 0 & \text{otherwise} \end{cases}$$

$$(4.3)$$

and

$$\phi_4^h(x) = \begin{cases} \dfrac{x - x_3}{\Delta x} & \text{for } x \in [x_3, x_4] \\ \dfrac{x_5 - x}{\Delta x} & \text{for } x \in]x_4, x_5] \\ 0 & \text{otherwise} \end{cases} \qquad \phi_5^h(x) = \begin{cases} \dfrac{x - x_4}{\Delta x} & \text{for } x \in [x_4, x_5] \\ 0 & \text{otherwise} \end{cases}$$

The set of a partition of the domain $\{K_i\}_{i=1,\dots,N_K}$, the set of degrees of freedom $\{x_j\}_{j=1,\dots,N_x}$, and a functional basis $\{\phi^h\}$ operating on an approximation space $\mathcal{V}_h$ defined according to $\{K_i\}$ and $\{x_j\}$ (referred to as $\mathbb{P}_1$ functions), are called a finite element representation.

To find an approximation solution u_h of the weak solution u of Eq. (4.2), assume a decomposition for u_h involving the functional basis $\{\phi^h\}$ of $\mathcal{V}_h$ and the weights $\bar{u}_i$:

$$u(x) \simeq u_h(x) = \sum \bar{u}_i \phi_i^h(x).$$

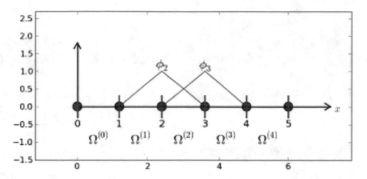

Fig. 4.1 Illustration of the *hat function*, i.e. $\mathbb{P}_1$−Lagrange test function. *Source* http://i.stack.imgur.com/Lt5Lw.png

Thus, the approximation of the weak solution u using the $\mathbb{P}_1$ finite element defined previously in $]a, b[$ becomes

$$u(x) \simeq \sum_{i=0}^{5} \bar{u}_i \phi_i^h(x)$$

$$\simeq \bar{u}_0 \phi_0^h(x) + \bar{u}_1 \phi_1^h(x) + \bar{u}_2 \phi_2^h(x) + \bar{u}_3 \phi_3^h(x) + \bar{u}_4 \phi_4^h(x) + \bar{u}_5 \phi_5^h(x).$$

By considering the Dirichlet boundary conditions, it is straightforward to conclude that

$$\bar{u}_0 = u_a, \qquad\qquad\qquad \bar{u}_5 = u_b.$$

Thus, only the unknown weights $\bar{u}_i$, $i = 1, 2, 3, 4$ must be computed.

Similarly, we define the approximation space $\mathscr{V}_{h,0}$ for the test functions $\{\phi_j^h\}$ that vanish at the boundaries. Recall that it is required that the test functions vanish at the boundary after the integration by parts in order to obtain the variational formulation of the problem. An obvious choice is to consider

$$\mathscr{V}_{h,0} = \{\phi^h \in \mathscr{V}_h \mid \phi^h(a) = 0, \phi^h(b) = 0\}.$$

In other words, in the present example, this space is spanned using the basis

$$\mathscr{V}_{h,0} = \text{Span} \{\phi_j^h\} \text{ for } j = 1, \ldots, 4$$

Remark 4.4 The choice of the same notation for the test functions occurring in the weak formulation and the basis elements is not coincidental: u_h and the test functions ϕ^h are considered in almost the same space.

Thus, the finite element formulation of the variational problem is as follows.

Find $u^h \in \mathscr{V}^h$ as the solution of

$$\int_a^b \left(\sum_{i=0}^5 \bar{u}_i \frac{\mathrm{d}\phi_i^h}{\mathrm{d}x}(x) \cdot \frac{\mathrm{d}\phi_j^h}{\mathrm{d}x}(x) \right) \mathrm{d}x = \int_a^b \left(f(x) \cdot \phi_j^h(x) \right) \mathrm{d}x \quad \forall \phi_j^h \in \mathscr{V}_{h,0}, \quad (4.4)$$

with $\bar{u}_0 = u_a$ and $\bar{u}_5 = u_b$.

4.1.3 Finite Element Approximation

The unknowns to be determined are $\bar{u}_1, \bar{u}_2, \bar{u}_3$, and $\bar{u}_4$. It is expected that the number of basis test functions in $\mathscr{V}_{h,0}$ is also four, corresponding to ϕ_j^h, $j = 1, \ldots, 4$, as defined in Eq. (4.3). Thus, the weak formulation obtained through the finite element representation in Eq. (4.4) yields four equalities to be verified by the four unknowns $\bar{u}_1, \bar{u}_2, \bar{u}_3$, and $\bar{u}_4$. As the support of each ϕ_j^h (i.e. the interval over which each ϕ_j^h does not vanish) is exactly $]x - j - 1, x_{j+1}[$, the integral in Eq. (4.4) can be reduced along with the number of terms occurring in the sum (corresponding to the u^h expansion).

Thus, the weak formulation, that is, Eq. (4.4), reads explicitly as follows.

- For ϕ_1^h,

$$\int_{x_0}^{x_1} \left(\left[\bar{u}_0 \frac{\mathrm{d}\phi_0^h}{\mathrm{d}x}(x) + \bar{u}_1 \frac{\mathrm{d}\phi_1^h}{\mathrm{d}x}(x) \right] \cdot \frac{\mathrm{d}\phi_1^h}{\mathrm{d}x}(x) \right) \mathrm{d}x +$$

$$\int_{x_1}^{x_2} \left(\left[\bar{u}_1 \frac{\mathrm{d}\phi_1^h}{\mathrm{d}x}(x) + \bar{u}_2 \frac{\mathrm{d}\phi_2^h}{\mathrm{d}x}(x) \right] \cdot \frac{\mathrm{d}\phi_1^h}{\mathrm{d}x}(x) \right) \mathrm{d}x = \int_{x_0}^{x_2} f(x) \cdot \phi_1^h(x) \, \mathrm{d}x.$$

- For ϕ_2^h,

$$\int_{x_1}^{x_2} \left(\left[\bar{u}_1 \frac{\mathrm{d}\phi_1^h}{\mathrm{d}x}(x) + \bar{u}_2 \frac{\mathrm{d}\phi_2^h}{\mathrm{d}x}(x) \right] \cdot \frac{\mathrm{d}\phi_2^h}{\mathrm{d}x}(x) \right) \mathrm{d}x +$$

$$\int_{x_2}^{x_3} \left(\left[\bar{u}_2 \frac{\mathrm{d}\phi_2^h}{\mathrm{d}x}(x) + \bar{u}_3 \frac{\mathrm{d}\phi_3^h}{\mathrm{d}x}(x) \right] \cdot \frac{\mathrm{d}\phi_2^h}{\mathrm{d}x}(x) \right) \mathrm{d}x = \int_{x_1}^{x_3} f(x) \cdot \phi_2^h(x) \, \mathrm{d}x.$$

- For ϕ_3^h,

$$\int_{x_2}^{x_3} \left(\left[\bar{u}_2 \frac{\mathrm{d}\phi_2^h}{\mathrm{d}x}(x) + \bar{u}_3 \frac{\mathrm{d}\phi_3^h}{\mathrm{d}x}(x) \right] \cdot \frac{\mathrm{d}\phi_3^h}{\mathrm{d}x}(x) \right) \mathrm{d}x +$$

$$\int_{x_3}^{x_4} \left(\left[\bar{u}_3 \frac{\mathrm{d}\phi_3^h}{\mathrm{d}x}(x) + \bar{u}_4 \frac{\mathrm{d}\phi_4^h}{\mathrm{d}x}(x) \right] \cdot \frac{\mathrm{d}\phi_3^h}{\mathrm{d}x}(x) \right) \mathrm{d}x = \int_{x_2}^{x_4} f(x) \cdot \phi_3^h(x) \, \mathrm{d}x.$$

- Finally, for ϕ_4^h,

$$\int_{x_3}^{x_4} \left(\left[\bar{u}_3 \frac{d\phi_3^h}{dx}(x) + \bar{u}_4 \frac{d\phi_4^h}{dx}(x) \right] \cdot \frac{d\phi_4^h}{dx}(x) \right) dx +$$

$$\int_{x_4}^{x_5} \left(\left[\bar{u}_4 \frac{d\phi_4^h}{dx}(x) + \bar{u}_5 \frac{d\phi_5^h}{dx}(x) \right] \cdot \frac{d\phi_4^h}{dx}(x) \right) dx = \int_{x_2}^{x_4} f(x) \cdot \phi_4^h(x) \, dx.$$

A compact formulation for $j = 1, \ldots, 4$ is simply written as

$$\int_{x_{j-1}}^{x_j} \left[\bar{u}_{j-1} \frac{d\phi_{j-1}^h}{dx} + \bar{u}_j \frac{d\phi_j^h}{dx} \right] \cdot \frac{d\phi_j^h}{dx} \, dx + \int_{x_j}^{x_{j+1}} \left[\bar{u}_j \frac{d\phi_j^h}{dx} + \bar{u}_{j+1} \frac{d\phi_{j+1}^h}{dx} \right] \cdot \frac{d\phi_j^h}{dx} \, dx$$

$$= \int_{x_{j-1}}^{x_{j+1}} f \cdot \phi_j^h \, dx.$$

Moreover, we obtain the following from an easy computation of the derivatives of each ϕ_j^h.

- For $x \in [x_{j-1}, x_j]$,

$$\frac{d\phi_{j-1}^h}{dx}(x) = \frac{-1}{\Delta x}, \qquad\qquad \frac{d\phi_j^h}{dx}(x) = \frac{+1}{\Delta x},$$

- For $x \in [x_j, x_{j+1}]$,

$$\frac{d\phi_j^h}{dx}(x) = \frac{-1}{\Delta x}, \qquad\qquad \frac{d\phi_{j+1}^h}{dx}(x) = \frac{+1}{\Delta x},$$

Thus, each variational problem is as follows for $j = 1, \ldots, 4$:

$$\int_{x_{j-1}}^{x_j} \left(-\frac{\bar{u}_{j-1}}{\Delta x^2} + \frac{\bar{u}_j}{\Delta x^2} \right) dx + \int_{x_j}^{x_{j+1}} \left(+\frac{\bar{u}_j}{\Delta x^2} - \frac{\bar{u}_{j+1}}{\Delta x^2} \right) dx = \int_{x_{j-1}}^{x_{j+1}} f \cdot \phi_j^h \, dx.$$

As each interval is of length Δx, then for $j = 1, \ldots, 4$,

$$-\frac{\bar{u}_{j-1}}{\Delta x} + 2\frac{\bar{u}_j}{\Delta x} - \frac{\bar{u}_{j+1}}{\Delta x} = \int_{x_{j-1}}^{x_{j+1}} f \cdot \phi_j^h \, dx.$$

Specifically, for $j = 1$, as the Dirichlet boundary condition imposes $\bar{u}_0 = u_a$,

$$2\bar{u}_1 - \bar{u}_2 = \Delta x \int_{x_0}^{x_2} f \cdot \phi_1^h \, dx + u_a.$$

For the following two equations, $j = 2$ and $j = 3$, respectively.

$$-\bar{u}_1 + 2\bar{u}_2 - \bar{u}_3 = \Delta x \int_{x_1}^{x_3} f \cdot \phi_2^h \, dx,$$

$$-\bar{u}_2 + 2\bar{u}_3 - \bar{u}_4 = \Delta x \int_{x_2}^{x_4} f \cdot \phi_3^h \, dx.$$

Furthermore, for $j = 4$, as the Dirichlet boundary condition is $\bar{u}_5 = u_b$,

$$-\bar{u}_3 + 2\bar{u}_4 = \Delta x \int_{x_3}^{x_5} f \cdot \phi_4^h \, dx + u_b.$$

Thus, the final system is

$$A\mathbf{u} = b,$$

with

$$A = \begin{pmatrix} 2 & -1 & 0 & 0 \\ -1 & 2 & -1 & 0 \\ 0 & -1 & 2 & -1 \\ 0 & 0 & -1 & 2 \end{pmatrix} \quad \text{and} \quad b = \begin{pmatrix} b_1 \\ b_2 \\ b_3 \\ b_4 \end{pmatrix} + \begin{pmatrix} u_a \\ 0 \\ 0 \\ u_b \end{pmatrix}$$

where $b_i = \Delta x \int_{x_{i-1}}^{x_{i+1}} f \cdot \phi_i^h$, $i = 1, 2, 3$. The integration of each b_i can be realized explicitly by using a quadrature formula.

Remark 4.5 Formally, this matricial system has the same structure as that obtained using a finite difference method. However, the concept involved here is relatively different.

4.2 Finite Element Approach Revisited

There is a more systematic (and more convenient method when considering a higher dimensional problem) to construct the final matrix system obtained using a finite element representation. It consists of first evaluating the contribution of each element K_i, and then easily satisfying the stiffness matrix by using a correspondence table.

4.2.1 Reference Element

In the computation of a weak formulation on each basis test function ϕ_j, the weak formulation may appear repetitive. The explicit formula of a test function could be unnecessarily tedious in case of a complex geometry, which uses the absolute coordinate of each point. Moreover, it uses the same generic formula. Instead, one can compute the change of variable for each segment K_i such that the computation of each element contribution is reduced to the contribution of a single generic element, using the change of variable given by the absolute value of a determinant of a Jacobian matrix $|J_{K_i}|$ associated to the change:

$$\int_{K_i} [\cdot] \, dx = \int_0^1 [\cdot] \, |J_{K_i}| \, d\xi.$$

In the reference element, the basis functions of the $\mathbb{P}_1$ test are simply given as

$$\psi_L(\xi) = 1 - \xi, \qquad\qquad \psi_R(\xi) = \xi.$$

Thus, their derivatives are trivially given as

$$\psi_L'(\xi) = -1, \qquad\qquad \psi_R'(\xi) = 1.$$

For example, consider the first strictly interior subdomain K_2:

- Recall that $K_2 = [x_1, x_2]$, where $x_1 = a + \Delta x$ and $x_2 = a + 2\Delta x$.
- The nonzero test functions on K_2 are

$$\phi_1(x) = \frac{x_2 - x}{\Delta x} \qquad \text{and} \qquad \phi_2(x) = \frac{x - x_1}{\Delta x}.$$

- Consider the change of variable such that the integral over K_2 becomes an integral over the unit segment:

$$\xi := \frac{x - x_1}{|K_2|},$$

and thus the test functions have a slope of ± 1. The determinant of the Jacobian J_K of this transformation is exactly K_2^{-1}.
- According to the formula of change of variables, the LHS of the variational formulation contributing to K_2 reads as

$$\int_{K_2} \left(\sum_{i=1}^{2} \bar{u}_i \frac{d\phi_i^h}{dx} \cdot \frac{d\phi_1^h}{dx} \right) dx = \int_0^1 \left(\sum_{i=1}^{2} \bar{u}_i \frac{d\psi_i^h}{d\xi} \cdot \frac{d\psi_1^h}{d\xi} \right) |J_{K_2}| \, d\xi,$$

and

$$\int_{K_2} \left(\sum_{i=1}^{2} \bar{u}_i \frac{d\phi_i^h}{dx}(x) \cdot \frac{d\phi_2^h}{dx}(x) \right) dx = \int_0^1 \left(\sum_{i=1}^{2} \bar{u}_i \frac{d\psi_i^h}{d\xi} \cdot \frac{d\psi_2^h}{d\xi} \right) |J_{K_2}| \, d\xi.$$

- A straightforward computation yields

$$\int_0^1 \left(\sum_{i=1}^2 \bar{u}_i \frac{\mathrm{d}\psi_i^h}{\mathrm{d}\xi} \cdot \frac{\mathrm{d}\psi_1^h}{\mathrm{d}\xi} \right) |J_{K_2}| \, \mathrm{d}\xi = \frac{+1}{|K_2|}\bar{u}_1 + \frac{-1}{|K_2|}\bar{u}_2,$$

and

$$\int_0^1 \left(\sum_{i=1}^2 \bar{u}_i \frac{\mathrm{d}\psi_i^h}{\mathrm{d}\xi} \cdot \frac{\mathrm{d}\psi_2^h}{\mathrm{d}\xi} \right) |J_{K_2}| \, \mathrm{d}\xi = \frac{-1}{|K_2|}\bar{u}_1 + \frac{+1}{|K_2|}\bar{u}_2.$$

Thus, the K_2 contribution to the stiffness matrix A_2 is

$$A_2 \begin{pmatrix} \bar{u}_1 \\ \bar{u}_2 \end{pmatrix} = \frac{1}{|K_2|} \begin{pmatrix} 1 & -1 \\ -1 & 1 \end{pmatrix} \begin{pmatrix} \bar{u}_1 \\ \bar{u}_2 \end{pmatrix}$$

Similarly,

• The K_3 contribution to the stiffness matrix A_3 is

$$A_3 \begin{pmatrix} \bar{u}_2 \\ \bar{u}_3 \end{pmatrix} = \frac{1}{|K_3|} \begin{pmatrix} 1 & -1 \\ -1 & 1 \end{pmatrix} \begin{pmatrix} \bar{u}_2 \\ \bar{u}_3 \end{pmatrix}$$

• and the K_4 contribution to the stiffness matrix A_4 is

$$A_4 \begin{pmatrix} \bar{u}_3 \\ \bar{u}_4 \end{pmatrix} = \frac{1}{|K_4|} \begin{pmatrix} 1 & -1 \\ -1 & 1 \end{pmatrix} \begin{pmatrix} \bar{u}_3 \\ \bar{u}_4 \end{pmatrix}$$

Now, consider the first subdomain, involving the Dirichlet boundary condition $u(a) = u_a$. This boundary condition imposes $\bar{u}_0 = u_a$, such that the boundary node is not an unknown. The K_1 contribution reads as

$$\int_{K_1} \left(\sum_{i=0}^1 \bar{u}_i \frac{\mathrm{d}\phi_i^h}{\mathrm{d}x} \cdot \frac{\mathrm{d}\phi_1^h}{\mathrm{d}x} \right) \mathrm{d}x = \int_0^1 \left(\sum_{i=0}^1 \bar{u}_i \frac{\mathrm{d}\psi_i^h}{\mathrm{d}\xi} \cdot \frac{\mathrm{d}\psi_1^h}{\mathrm{d}\xi} \right) |J_{K_1}| \, \mathrm{d}\xi.$$

Recall that we do not have to compute the variational formulation on ϕ_0 because $\phi_0 \notin \mathcal{V}_{h,0}$; however, we must consider it when decomposing $u_h \in \mathcal{V}_h$. Finally, the K_1 contribution, A_1, is given by the following vector product:

$$A_1 \begin{pmatrix} u_a \\ \bar{u}_1 \end{pmatrix} = \frac{1}{|K_1|} (-1 \ 1) \begin{pmatrix} u_a \\ \bar{u}_1 \end{pmatrix}$$

The term containing the u_a node must be moved to the RHS of the final matrix system. The other Dirichlet boundary condition $\bar{u}_5 = u_b$ is similarly dealt with to yield to the K_5 contribution, A_5, which is given by the scalar relation

$$A_5 \begin{pmatrix} \bar{u}_4 \\ u_b \end{pmatrix} = \frac{1}{|K_5|} (-1 \ 1) \begin{pmatrix} \bar{u}_4 \\ u_b \end{pmatrix}.$$

4.2.2 Connectivity Table

The connectivity table is generated using a mesh generator software (e.g.gmsh). It relies on the label number of each degree of freedom, its neighbor, and the element K_i that it belongs to, so that the reconstruction of the global stiffness matrix composed by all the contributions of each subdomain can systematically be accomplished. Here, for each K_i, the connectivity table would simply indicate the degree of freedom corresponding to the left boundary (mapped to zero when the variable is changed to the reference segment) and the right boundary (mapped to one when the variable is changed to the reference segment).

4.2.3 Stiffness Matrix Construction

The assembly of all the contributions of $A_1, A_2, \ldots, A_5$ computed previously, yields the final matrix system:

$$A\mathbf{u} = b,$$

with $|K_i| = \Delta x, i = 1, \ldots, 5,$

$$A = \frac{1}{\Delta x}\begin{pmatrix} 1+1 & -1 & 0 & 0 \\ -1 & 1+1 & -1 & 0 \\ 0 & -1 & 1+1 & 1 \\ 0 & 0 & -1 & 1+1 \end{pmatrix} \quad \text{and} \quad b = \begin{pmatrix} b_1 \\ b_2 \\ b_3 \\ b_4 \end{pmatrix} + \frac{1}{\Delta x}\begin{pmatrix} u_a \\ 0 \\ 0 \\ u_b \end{pmatrix},$$

where each b_i corresponds to the contribution of the RHS of the variational formulation. In addition, the contribution of each subdomain is highlighted using different colors

- K_1 in orange
- K_2 in red
- K_3 in cyan
- K_4 in brown
- K_5 in green

It is then easy to recover the matrix system computed in Sect. 4.1.3.

4.2.4 Final Remarks

The FEM is still largely used in engineering processes for numerical simulation. Many industrial software adopt this approach, taking advantage of its robustness, flexibility, adaptability to any type of geometry, and its capacity to simulate multiphysics problems. Theoretically, numerical analysis is also well established: the numerical properties are well understood, conferring a high level of confidence to this approach. Thus, compared with the finite difference method, the FEM proposes relatively more powerful aspects; however, the main distinctions arise in both the understanding of abstract concepts and the required computational techniques, at least for multidimensional cases. Currently, in most software dedicated to building physics, diffusion phenomena are modeled in 1D space. However, the recent developments in software that take into account more realistic effects (e.g. solar radiation, surrounding effects, and shadows) and involve advanced techniques (such as Delaunay triangulation and pixel counting) require transfer modeling that is more advanced than the modeling methods being currently used; the inclusion of generic geometries and spatial heterogeneities, for example, are parameters of great importance and are better supported by the FEM than by the finite difference method.

Part II
Advanced Numerical Methods

Chapter 5
Explicit Schemes with Improved CFL Condition

Above in Chap. 3 the basic finite differences approaches were presented. In particular, it was shown that explicit discretizations are subject to some additional constraints if one wants to have a stable[1] numerical scheme. These restrictions are known in the literature under the name of Courant–Friedrichs–Lewy conditions [43]. For parabolic diffusion equations they can be too prohibitive to use explicit schemes in general. In the present chapter we are going to present some strategies to relax (if not to remove completely) these limitations. The goal being to keep the simplicity of explicit schemes with wider stability region.

In the present chapter we consider the classical linear heat equation:

$$u_t = v \, \nabla^2 u \,, \tag{5.1}$$

where $u(x, t)$ is a quantity being diffused in some domain $\Omega \subseteq \mathbb{R}^d$. In physical applications $u(x, t), x \in \Omega, t > 0$ may represent the temperature field, moisture content, vapor concentration, etc. and $v > 0$ is the diffusion coefficient. The subscripts denote partial derivatives, i.e. $u_t \overset{\text{def}}{:=} \dfrac{\partial u(x, t)}{\partial t}$. Finally, $\nabla^2 \equiv \nabla \cdot \nabla$ is the classical $d-$dimensional Laplace operator:

$$\nabla^2 \overset{\text{def}}{:=} \sum_{i=1}^{d} \frac{\partial^2}{\partial x_i^2} \,.$$

The derivation of this equation for Brownian motion process was given by Einstein [58] in 1905.

---

[1]Notice, that stability does not imply the accuracy.

© Springer Nature Switzerland AG 2019
N. Mendes et al., *Numerical Methods for Diffusion Phenomena in Building Physics*,
https://doi.org/10.1007/978-3-030-31574-0_5

From now on we shall restrict our ambitions on the $1-$dimensional case where $\Omega \equiv [0, \ell] \subseteq \mathbb{R}^1$ and Eq. (5.1) correspondingly becomes:

$$u_t = \nu u_{xx}. \tag{5.2}$$

This equation has to be supplemented by one initial

$$u\big|_{t=0} = u_0(x),$$

and two boundary conditions:

$$\Phi_1\big(t, u(t, 0), u_x(t, 0)\big) = 0, \tag{5.3}$$

$$\Phi_r\big(t, u(t, \ell), u_x(t, \ell)\big) = 0. \tag{5.4}$$

The functions $\Phi_{1,r}(\bullet)$ have to be specified depending on the practical situation in hands. For example, if we have the Dirichlet-type condition on the left boundary then

$$\Phi_1\big(t, u(t, 0), u_x(t, 0)\big) \equiv u(t, 0) - u^\circ(t) = 0,$$

where $u^\circ(t)$ is a prescribed function of time. Often, it is assumed that $u^\circ(t) \equiv$ const. The homogeneous Neumann-type condition on the right looks like

$$\Phi_r\big(t, u(t, \ell), u_x(t, \ell)\big) \equiv u_x(t, \ell) = 0.$$

5.1   Some Healthy Criticism

Obviously, the heat equation (5.1) is a simplified model obtained after a series of idealizations and simplifications. As a result, Eq. (5.1) is linear and its Green function[2] can be computed analytically

$$G(x, t) = \frac{1}{\sqrt{4\pi \nu t}}\, e^{-\frac{x^2}{4\nu t}}.$$

In particular, one can see that for any sufficiently small $t > 0$ the function $G(x, t)$ is not of *compact support*. In other words, the information about a point source initially

---

[2]The Green function is also known as the fundamental solution. More precisely, it solves the following problem:

$$G_t = \nu G_{xx}, \qquad x \in \mathbb{R},$$

$$G\big|_{t=0} = \delta(x),$$

where $\delta(x)$ is Dirac delta function.

localized at $x = 0$ spreads instantly over the whole domain. Of course, the infinite speed of information propagation is physically forbidden. This non-physical feature of heat equation solutions is a consequence of simplifying assumptions made during the derivation. We just mention here that some nonlinear versions of the heat equation do have fundamental solutions with compact support.

However, in practice the solutions to Partial Differential Equations (PDEs) such as (5.1) are computed numerically than constructed analytically. That is why we can hope to correct some non-physical features of the heat equation solution at the discrete level. This is the main topic of the present chapter.

Chapter organization.

Below we shall review some classical numerical schemes for the heat equation (5.1) in one spatial dimension in Sect. 5.2:

- The explicit scheme in Sect. 5.2.1
- The implicit scheme in Sect. 5.2.2
- The leap-frog scheme in Sect. 5.2.3
- The Crank–Nicolson scheme in Sect. 5.2.4.

In that section we justify our preference for explicit schemes in time. However, explicit schemes are known to have an important CFL-type stability restrictions on the time step [43]. That is why in Sect. 5.3 we review some not so widely known schemes which allow to overcome the stability limit while still being explicit in time. In particular, we consider the following alternatives:

- Dufort–Frankel method in Sect. 5.3.1
- Saulyev method in Sect. 5.3.2
- Hyperbolization method in Sect. 5.3.3.

Finally, the main conclusions and perspectives of the present study are discussed in Sect. 5.4.

5.2 Classical Numerical Schemes

In order to describe numerical schemes in simple terms, consider a uniform discretization of the interval $\Omega \rightsquigarrow \Omega_h$:

$$\Omega_h = \bigcup_{j=0}^{N-1} [x_j, x_{j+1}], \qquad x_{j+1} - x_j \equiv \Delta x, \quad \forall j \in \{0, 1, \ldots, N-1\}.$$

The time layers are uniformly spaced as well $t^n = n \Delta t$, $\Delta t = \text{const} > 0$, $n = 0, 1, 2, \ldots$ The values of function $u(x, t)$ in discrete nodes will be denoted by $u_j^n \overset{\text{def}}{:=} u(x_j, t^n)$. The space-time grid is schematically depicted in Fig. 5.1.

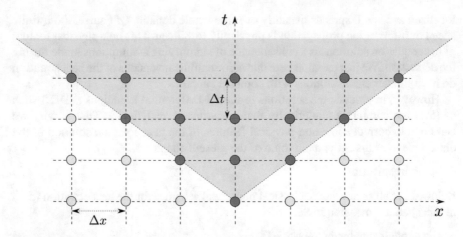

Fig. 5.1 A schematic representation of the uniform discretization in space and time. Red nodes correspond to non-zero values of the discrete solution. Grey nodes correspond to $u_j^n \equiv 0$. The shaded area is an equivalent of the '*light cone*' for the initially activated node

5.2.1 The Explicit Scheme

The standard explicit scheme for the linear heat equation (5.2) can be written as

$$\frac{u_j^{n+1} - u_j^n}{\Delta t} = \nu \frac{u_{j-1}^n - 2u_j^n + u_{j+1}^n}{\Delta x^2}, \qquad j = 1, \ldots, N-1, \qquad n \geqslant 0. \tag{5.5}$$

The stencil of this scheme is depicted in Fig. 5.2. In order to complete this discretization we have to find from boundary conditions (5.3), (5.4) the boundary values:

$$u_0^{n+1} = \psi_1(t^{n+1}, u_1^{n+1}, \ldots), \qquad u_N^{n+1} = \psi_r(t^{n+1}, u_{N-1}^{n+1}, \ldots), \tag{5.6}$$

where functions $\psi_{1,r}(\bullet)$ may depend on adjacent values of the solution whose number depends on the approximation order of the scheme (here we use the second order in space). For example, if the temperature is prescribed on the right boundary, then we simply have

$$u_N^{n+1} = \psi_r(t^{n+1}) \equiv \phi_r(t^{n+1}),$$

where $\phi_r(t)$ is a given function of time. On the other hand, if the heat flux is prescribed on the left boundary $\nu \dfrac{\partial u}{\partial x} = \phi_1(t)$ then it can be discretized as

$$\nu \frac{-3u_0^{n+1} + 4u_1^{n+1} - u_2^{n+1}}{2\Delta x} = \phi_1(t^{n+1}).$$

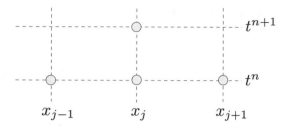

Fig. 5.2 Stencil of the explicit finite difference scheme (5.5)

By solving algebraic equation (5.5) with respect to u_j^{n+1} we obtain a discrete dynamical system

$$u_j^{n+1} = u_j^n + v \frac{\Delta t}{\Delta x^2} \left(u_{j-1}^n - 2 u_j^n + u_{j+1}^n \right),$$

whose starting value is directly obtained from the initial condition:

$$u_j^0 = u_0(x_j), \qquad j = 0, 1, \ldots, N.$$

It is well-known that scheme (5.5) approximates the continuous operator to order $\mathcal{O}(\Delta t + \Delta x^2)$. The explicit scheme is conditionally stable under the following CFL-type condition:

$$\Delta t \leqslant \frac{1}{2 v} \Delta x^2. \tag{5.7}$$

Unfortunately, this condition is too restrictive for sufficiently fine discretizations.

5.2.2   The Implicit Scheme

The implicit scheme for the 1D heat equation (5.2) is given by the following relations:

$$\frac{u_j^{n+1} - u_j^n}{\Delta t} = v \frac{u_{j-1}^{n+1} - 2 u_j^{n+1} + u_{j+1}^{n+1}}{\Delta x^2}, \qquad j = 1, \ldots, N-1, \qquad n \geqslant 0. \tag{5.8}$$

The finite difference stencil of this scheme is depicted in Fig. 5.3. These relations have to be properly initialized and supplemented with numerical boundary conditions (5.6). In the following sections we shall not return to the question of initial and boundary conditions in order to focus on the discretization. The scheme (5.8) has the same order of accuracy as the explicit scheme (5.5), i.e. $\mathcal{O}(\Delta t + \Delta x^2)$. However, the implicit scheme (5.8) is unconditionally stable, which constitutes its major advantage. It could be interesting to have also the second order in time as well. This issue will be addressed in the following sections.

Fig. 5.3 Stencil of the
implicit finite difference
scheme (5.8)

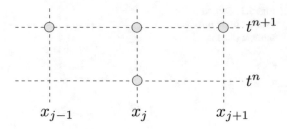

Fig. 5.4 Stencil of the
leap-frog (5.9) and
hyperbolic (5.23) finite
difference schemes

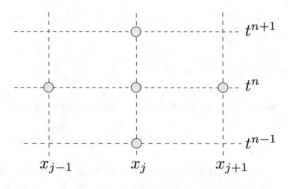

The most important difference with the explicit scheme (5.5) is that we have to solve a tridiagonal system of linear algebraic equations to determine the numerical solution values $\left\{u_j^{n+1}\right\}_{j=0}^{N}$ on the following time layer $t = t^{n+1}$. It determines the algorithm complexity—a tridiagonal system of equations can be solved in $\mathcal{O}(N)$ operations (using the simple Thomas algorithm, for example) and it has to be done at every time step.

5.2.3 The Leap-Frog Scheme

The leap-frog scheme[3], whose stencil is depicted in (Fig. 5.4), is obtained by replacing in (5.5) the forward difference in time by the symmetric one, i.e.

$$\frac{u_j^{n+1} - u_j^{n-1}}{2\,\Delta t} = \nu\,\frac{u_{j-1}^n - 2\,u_j^n + u_{j+1}^n}{\Delta x^2}, \qquad j = 1, \ldots, N-1, \qquad n \geqslant 0.$$
$$(5.9)$$

This scheme is second order accurate in space and in time, i.e. $\mathcal{O}(\Delta t^2 + \Delta x^2)$. Unfortunately, the leap-frog scheme is unconditionally unstable. It makes it un-exploitable in practice. However, we shall use some modifications of this scheme as shown in (Fig. 5.4).

---

[3]This scheme is called in French as '*le schéma saute-mouton*'.

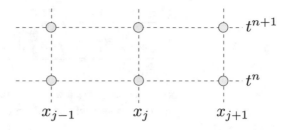

Fig. 5.5 Stencil of the Crank–Nicolson (CN) finite difference scheme (5.10)

5.2.4 The Crank–Nicolson Scheme

We saw above that the first tentative to obtain a scheme with second order accuracy in space *and* in time was unsuccessful (see Sect. 5.2.3). However, a very useful method was proposed by Crank and Nicolson (CN) and it can be successfully applied to the heat equation (5.2) as well:

$$\frac{u_j^{n+1} - u_j^n}{\Delta t} = \nu \, \frac{u_{j-1}^n - 2u_j^n + u_{j+1}^n}{2\,\Delta x^2} + \nu \, \frac{u_{j-1}^{n+1} - 2u_j^{n+1} + u_{j+1}^{n+1}}{2\,\Delta x^2},$$
$$j = 1, \ldots, N-1, \quad n \geqslant 0. \tag{5.10}$$

This scheme is $\mathcal{O}(\Delta t^2 + \Delta x^2)$ accurate and unconditionally stable (similarly to (5.8)). That is why numerical results obtained with the CN scheme will be more accurate than implicit scheme (5.8) predictions. The stencil of this scheme is depicted in Fig. 5.5. The CN scheme has all advantages and disadvantages (except for the order of accuracy in time) of the implicit scheme (5.8). At every time step one has to use a tridiagonal solver to invert the linear system of equations to determine solution value at the following time layer $t = t^{n+1}$.

5.2.4.1 Some Nonlinear Extensions

Since most of real-world heat conduction models used in building physics are nonlinear, it is worth to discuss some nonlinear extensions of the Crank–Nicolson (CN) scheme. For linear problems CN scheme turns out to be the same as the mid-point and trapezoidal rules for Ordinary Differential Equations (ODEs). Indeed, consider a nonlinear ODE:

$$\dot{u} = f(u), \quad u(0) = u_0. \tag{5.11}$$

The mid-point and trapezoidal rules consist correspondingly in discretizing (5.11) as follows:

$$\frac{u^{n+1} - u^n}{\Delta t} = f\left(\frac{u^n + u^{n+1}}{2}\right),$$

$$\frac{u^{n+1} - u^n}{\Delta t} = \frac{f(u^n) + f(u^{n+1})}{2}.$$

Now, if we set in formulas above $f(u) = v\mathcal{L} \cdot u$, where $\mathcal{L} \simeq \partial_{xx}$ is the linear operator which represents the second central finite difference, we recover the CN scheme (5.10).

Consider a non-conservative nonlinear heat equation:

$$u_t = k(u)\, u_{xx}. \tag{5.12}$$

The straightforward application of the CN scheme to Eq. (5.1) yields the following scheme:

$$\frac{u_j^{n+1} - u_j^n}{\Delta t} = k(u_j^n)\,\frac{u_{j-1}^n - 2u_j^n + u_{j+1}^n}{2\,\Delta x^2} + k(u_j^{n+1})\,\frac{u_{j-1}^{n+1} - 2u_j^{n+1} + u_{j+1}^{n+1}}{2\,\Delta x^2},$$

$$j = 1, \ldots, N-1, \qquad n \geqslant 0. \tag{5.13}$$

However, it is less known that one can apply also the cross-Crank–Nicolson (cCN) scheme:

$$\frac{u_j^{n+1} - u_j^n}{\Delta t} = k(u_j^{n+1})\,\frac{u_{j-1}^n - 2u_j^n + u_{j+1}^n}{2\,\Delta x^2} + k(u_j^n)\,\frac{u_{j-1}^{n+1} - 2u_j^{n+1} + u_{j+1}^{n+1}}{2\,\Delta x^2},$$

$$j = 1, \ldots, N-1, \qquad n \geqslant 0. \tag{5.14}$$

We underline that both schemes (5.13) and (5.14) are second order accurate in space *and* in time, i.e. the consistency error is $\mathcal{O}(\Delta t^2 + \Delta x^2)$. However, there is a major advantage of the cCN scheme (5.14) over the classical CN scheme (5.13) in the fact that cCN is linear with respect to quantities evaluated at the upcoming time layer $t = t^{n+1}$ provided that $k(u)$ is an affine function of u. This fact can be used to simplify the resolution procedure without destroying the accuracy of the CN scheme. Otherwise, for more general diffusion coefficients $k(u)$ the success of operation depends on the easiness to solve nonlinear equation (5.14). It goes without saying that information propagates instantaneously in both CN and cCN schemes.

5.2.5 Information Propagation Speed

Let us discuss now an important issue of the information propagation speed in the discretized heat equation (5.2). As the initial condition we take the following grid function:

$$u^0_j = \begin{cases} 1, & j = 0, \\ 0, & j \neq 0, \end{cases}$$

which corresponds to the discrete Dirac function. In all fully implicit schemes (such as (5.8) and (5.10)) the grid function $\{u^1_j\}^N_{j=0}$ will generally have non-zero values in all nodes (modulo perhaps homogeneous boundary conditions). Thus, we can conclude that information has spreaded instantaneously. On the other hand, as it is illustrated in Fig. 5.1 with grey and red circles, in explicit discretizations the information propagates one cell to the left and one cell to the right in one time step. Thus, its speed c_s can be estimated as

$$c_s = \frac{\Delta x}{\Delta t} \underset{\text{CFL}}{\geqslant} \frac{2\nu}{\Delta x}.$$

Thus, the value of c_s is finite. Of course, in the limit $\Delta x \to 0$ we recover the infinite propagation speed, but let us not forget that computations are always run for a *finite* value of Δx.

We arrived to an interesting conclusion. Even if the continuous heat equation (5.1) possesses an unphysical property, it can be corrected if we use a judicious (in this case *explicit*) discretization. This is the main reason why we privilege explicit schemes in time. However, these schemes are subject to severe stability restrictions. The rest of the chapter is devoted to the question how to overcome the stability limit?

There is another computational advantage of explicit schemes over the implicit ones. Namely, explicit methods can be easily parallelized and they allow to achieve almost perfect scaling on HPC systems [35]. Indeed, the computational domain can be split into sub-domains, each sub-domain being handled by a separate processor. Since the stencil is local, only direct neighbours are involved in individual computations. The communication among various processes is almost minimal since only boundary nodes have to be shared. This is another good reason to privilege explicit schemes over the implicit ones.

5.3 Improved Explicit Schemes

Below we present several alternative methods which were specifically designed to overcome the stability limitation of the standard explicit scheme (5.5).

5.3.1 Dufort–Frankel Method

Let us take the unconditionally unstable leap-frog scheme (5.9) and slightly modify it to obtain the so-called Dufort–Frankel method:

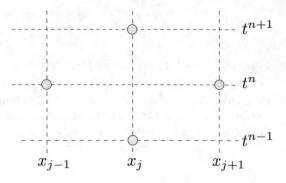

Fig. 5.6 Stencil of the Dufort–Frankel (5.15) finite difference scheme

$$\frac{u_j^{n+1} - u_j^{n-1}}{2\,\Delta t} = \nu\,\frac{u_{j-1}^n - \left(u_j^{n-1} + u_j^{n+1}\right) + u_{j+1}^n}{\Delta x^2},$$
$$j = 1, \ldots, N-1, \quad n \geqslant 0, \tag{5.15}$$

where we made a replacement

$$2\,u_j^n \hookleftarrow u_j^{n-1} + u_j^{n+1}.$$

The scheme (5.15) has the stencil depicted in Fig. 5.6. At the first glance the scheme (5.15) looks like an implicit scheme, however, it is not truly the case. Equation (5.15) can be easily solved for u_j^{n+1} to give the following discrete dynamical system:

$$u_j^{n+1} = \frac{1 - \lambda}{1 + \lambda}\,u_j^{n-1} + \frac{\lambda}{1 + \lambda}\left(u_{j+1}^n + u_{j-1}^n\right),$$

where

$$\lambda \stackrel{\text{def}}{:=} 2\,\nu\,\frac{\Delta t}{\Delta x^2}.$$

The standard von Neumann stability analysis shows that the Dufort–Frankel scheme is *unconditionally stable*.

The consistency error analysis of the scheme (5.15) shows the following interesting result:

$$\mathcal{L}_j^n = \underbrace{\nu\,\frac{\Delta t^2}{\Delta x^2}\,u_{tt}}_{\equiv\,\tau} + \underbrace{u_t - \nu\,u_{xx}}_{(5.2)} +$$

$$\tfrac{1}{6}\,\Delta t^2\,u_{ttt} - \tfrac{1}{12}\,\nu\,\Delta x^2\,u_{xxxx} - \tfrac{1}{12}\,\nu\,\Delta t^2\,\Delta x\,u_{xxxtt} + \mathcal{O}\!\left(\frac{\Delta t^4}{\Delta x^2}\right),$$

where

$$\mathcal{L}_j^n \overset{\text{def}}{:=} \frac{u_j^{n+1} - u_j^{n-1}}{2\,\Delta t} - \nu\, \frac{u_{j-1}^n - \left(u_j^{n-1} + u_j^{n+1}\right) + u_{j+1}^n}{\Delta x^2}.$$

So, from the asymptotic expansion for $\mathcal{L}_j^n$ we obtain that the Dufort–Frankel scheme is second order accurate in time *and*

- First order accurate in space if $\Delta t \propto \Delta x^{3/2}$
- Second order accurate in space if $\Delta t \propto \Delta x^2$

However, the Dufort–Frankel scheme is unconditionally consistent with the so-called *hyperbolic heat conduction equation*:

$$\tau\, u_{tt} + u_t - \nu\, u_{xx} = 0.$$

We shall return to this equation below. At this stage we only mention that information propagates with the finite speed in hyperbolic models.

5.3.2 Saulyev Method

In this section we describe a not so widely known method proposed by Saulyev in [175] to integrate parabolic equations. For simplicity we focus on the 1−dimensional heat equation (5.2). The first idea of this method consists in rewriting the discrete second (spatial) derivative as

$$u_{xx}\big|_{x=x_j} \approx \frac{u_{j+1} - 2u_j + u_{j-1}}{\Delta x^2} \equiv \frac{\dfrac{u_{j+1} - u_j}{\Delta x} - \dfrac{u_j - u_{j-1}}{\Delta x}}{\Delta x}.$$

In the finite difference formula above we do not specify intentionally the time layer number. The next trick consists in writing the following asymmetric finite difference approximation:

$$\frac{u_j^{n+1} - u_j^n}{\Delta t} = \nu\, \frac{\dfrac{u_{j+1}^n - u_j^n}{\Delta x} - \dfrac{u_j^{n+1} - u_{j-1}^{n+1}}{\Delta x}}{\Delta x},$$

or after simplifications we obtain

$$\frac{u_j^{n+1} - u_j^n}{\Delta t} = \nu\, \frac{u_{j+1}^n - \left(u_j^n + u_j^{n+1}\right) + u_{j-1}^{n+1}}{\Delta x^2}. \tag{5.16}$$

The last difference relation is slightly different from the Dufort–Frankel method presented above. Moreover, the relation written above is not consistent with the original equation (5.2). That is why we consider the next time layer $t = t^{n+2}$ and we apply symmetrically the same tricks, i.e.

$$\frac{u_j^{n+2} - u_j^{n+1}}{\Delta t} = \nu \frac{\dfrac{u_{j+1}^{n+2} - u_j^{n+2}}{\Delta x} - \dfrac{u_j^{n+1} - u_{j-1}^{n+1}}{\Delta x}}{\Delta x},$$

and after simplifications we have

$$\frac{u_j^{n+2} - u_j^{n+1}}{\Delta t} = \nu \frac{u_{j+1}^{n+2} - \left(u_j^{n+2} + u_j^{n+1}\right) + u_{j-1}^{n+1}}{\Delta x^2}. \tag{5.17}$$

Without any surprise the relation (5.17) does not approximate equation (5.2) either. However, both relations (5.16) and (5.17) constitute the so-called Saulyev scheme and are called the first and second stages of Saulyev method correspondingly.

In order to perform the approximation error analysis we take the sum of (5.16), (5.17) and we divide it by two:

$$\mathscr{L}_j^n = \frac{u_j^{n+2} - u_j^n}{2\,\Delta t} = \nu \frac{u_j^n - u_{j+1}^n - 2u_{j-1}^{n+1} + 2u_j^{n+1} + u_j^{n+2} - u_{j+1}^{n+2}}{2\,\Delta x^2}.$$

After applying local Taylor expansions we obtain

$$\mathscr{L}_j^n = u_t - \nu u_{xx} - \tfrac{1}{12}\nu\,\Delta x^2\,u_{xxxx} +$$
$$\left[\tfrac{2}{3}u_{ttt} - \tfrac{3}{4}\nu u_{xxtt}\right]\Delta t^2 - \tfrac{1}{2}\nu\frac{\Delta t^2}{\Delta x}u_{xtt} + \mathscr{O}(\Delta t\,\Delta x^2 + \Delta t^2\,\Delta x).$$

From the last asymptotic expansion we arrive at an important result: Saulyev scheme is second order accurate in space if $\Delta t = \mathscr{O}(\Delta x^{3/2})$. We underline the fact that this condition coming from the accuracy requirements is weaker than the usual CFL restriction (5.7). For instance, if the user is ready to sacrifice the spatial accuracy to the first order $\mathscr{O}(\Delta x)$, then it is sufficient to take $\Delta t = \mathscr{O}(\Delta x)$.

Without proof we report that Saulyev's scheme is unconditionally stable (see [175] for more details). The stencil of Saulyev's scheme is depicted in Fig. 5.7.

5.3.2.1   Resolution Procedure

At the first glance Saulyev scheme appears as an implicit one since each relation (5.16) and (5.17) contains two terms from the following time layer ($t = t^{n+1}$ and $t = t^{n+2}$ correspondingly). However, this scheme can be recast in an almost explicit form using judicious recurrence relations.

Consider the first stage (5.16) of Saulyev's scheme. Similarly to Sect. 5.3.1 we introduce for simplicity the parameter

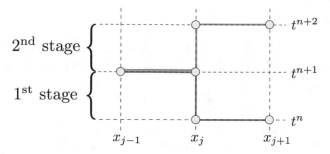

Fig. 5.7 Stencil of the Saulyev finite difference scheme, which consists of two stages (5.16) and (5.17)

$$\lambda \overset{\text{def}}{:=} \nu \, \frac{\Delta t}{\Delta x^2} \, .$$

From difference relation (5.16) we find

$$u_j^{n+1} \;=\; \frac{1-\lambda}{1+\lambda} \, u_j^n \;+\; \frac{\lambda}{1+\lambda} \, u_{j+1}^n \;+\; \frac{\lambda}{1+\lambda} \, u_{j-1}^{n+1} . \qquad (5.18)$$

The first stage of Saulyev's scheme is computed in rightwards direction (increasing index $j \nearrow$). From the left boundary condition we find first the value

$$u_0^{n+1} \;=\; \psi_1 \, (t^{n+1}) \, .$$

This allows us to compute u_1^{n+1}, u_2^{n+1}, etc. thanks to the recurrence relation (5.18). At the final step, the value u_N^{n+1} is computed directly from the right boundary condition:

$$u_N^{n+1} \;=\; \psi_{\text{r}} \, (t^{n+1}) \, .$$

This completes the first stage of computations.

Remark 5.1 For some types of boundary conditions (e.g. Robin-type) Saulyev's scheme might require solution of a small dimensional (typically 2×2) linear system of algebraic equations to initiate the recurrence (5.18).

Let us make explicit now the second stage (5.17) of Saulyev's method. For this purpose we solve relation (5.17) with respect to u_j^{n+2}:

$$u_j^{n+2} \;=\; \frac{1-\lambda}{1+\lambda} \, u_j^{n+1} \;+\; \frac{\lambda}{1+\lambda} \, u_{j-1}^{n+1} \;+\; \frac{\lambda}{1+\lambda} \, u_{j+1}^{n+2} . \qquad (5.19)$$

Now it is getting clear that during the second stage of Saulyev's scheme we proceed in the leftwards direction (decreasing $j \searrow$). From the right boundary condition we find first

$$u_N^{n+2} = \psi_r(t^{n+2}).$$

It allows us to compute u_{N-1}^{n+2}, u_{N-2}^{n+2}, etc. using the recurrence relation (5.19). At the final step, the value u_0^{n+2} is computed directly from the left boundary condition:

$$u_0^{n+2} = \psi_1(t^{n+2}).$$

As a result we obtain a fully explicit resolution scheme without stability related limitations. We notice however that Saulyev's scheme provides consistent results only every second time step or, in other words, after the successive completion of both stages (5.18) and (5.19). The intermediate result is not consistent with the equation (5.2).

5.3.3 Hyperbolization Method

We saw above that the Dufort–Frankel scheme is a hidden way to add a small amount of 'hyperbolicity' into the model (5.2). In this section we shall invert the order of operations: first, we perturb the equation (5.2) in an ad-hoc way and only after we discretize it with a suitable method.

Consider the 1−dimensional heat equation (5.2) that we are going to perturb by adding a small term containing the second derivative in time:

$$\tau u_{tt} + u_t - \nu u_{xx} = 0. \tag{5.20}$$

This is the *hyperbolic heat equation* already familiar to us since it appeared in the consistency analysis of the Dufort–Frankel scheme. Here we perform a singular perturbation by assuming that

$$\| \tau u_{tt} \| \ll \| u_t \|.$$

The last condition physically means that the new term has only limited influence on the solution of equation (5.20). Here τ is a small ad-hoc parameter whose value is in general related to physical and discretization parameters $\tau = \tau(\nu, \Delta x, \Delta t)$.

Remark 5.2 One can notice that Eq. (5.20) is second order in time, thus, it requires two initial conditions to obtain a well-posed initial value problem. However, the parabolic equation (5.2) is only first order in time and it only requires the knowledge of the initial temperature field distribution. When we solve the hyperbolic equation (5.20), the missing initial condition is simply chosen to be

$$u_t\big|_{t=0} = 0.$$

5.3.3.1 Dispersion Relation Analysis

The classical dispersion relation analysis looks at plane wave solutions:

$$u(x, t) = u_0 e^{i(\kappa x - \omega t)}. \tag{5.21}$$

By substituting this solution ansatz into Eq. (5.2) we obtain the following relation between wave frequency ω and wavenumber k:

$$\omega(\kappa) = -i \nu \kappa^2. \tag{5.22}$$

The last relation is called the *dispersion relation* even if the heat equation (5.2) is not dispersive but dissipative. The real part of ω contains information about wave propagation properties (dispersive if $\frac{\text{Re } \omega(\kappa)}{\kappa} \neq$ const and non-dispersive otherwise) while the imaginary part describes how different modes κ dissipate (if $\text{Im } \omega(\kappa) < 0$) or grow (if $\text{Im } \omega(\kappa) > 0$). The dispersion relation (5.22) gives the damping rate of different modes.

The same plane wave ansatz (5.21) can be substituted into the hyperbolic heat equation (5.20) as well to give the following *implicit* relation for the wave frequency ω:

$$-\tau \omega^2 - i\omega + \nu \kappa^2 = 0.$$

By solving this quadratic equation with complex coefficients for ω, we obtain two branches:

$$\omega_{\pm}(\kappa) = \frac{-i \pm \sqrt{4 \nu \kappa^2 \tau - 1}}{2\tau}.$$

This dispersion relation will be analyzed asymptotically with $\tau \ll 1$ being the small parameter. The branch $\omega_-(\kappa)$ is not of much interest to us since it is constantly damped, i.e.

$$\omega_-(\kappa) = -\frac{i}{\tau} + \mathcal{O}(1).$$

It is much more instructive to look at the positive branch $\omega_+(\kappa)$:

$$\omega_+(\kappa) = -i \nu \kappa^2 \left[1 + \nu \kappa^2 \tau + 2 \nu^2 \kappa^4 \tau^2 + \mathcal{O}(\tau^3) \right].$$

The last asymptotic expansion shows that for small values of parameter τ we obtain a valid asymptotic approximation of the dispersion relation (5.22) for the heat equation (5.2).

5.3.3.2 Discretization

Equation (5.20) will be discretized on the same stencil as the leap-frog scheme (5.9) (see Fig. 5.4):

$$\mathscr{L}_j^n \stackrel{def}{:=} \tau \frac{u_j^{n+1} - 2u_j^n + u_j^{n-1}}{\Delta t^2} + \frac{u_j^{n+1} - u_j^{n-1}}{2\Delta t} - \nu \frac{u_{j+1}^n - 2u_j^n + u_{j-1}^n}{\Delta x^2} = 0,$$

$$j = 1, \ldots, N - 1, \qquad n \geqslant 0,$$
$$(5.23)$$

The last scheme is consistent with hyperbolic heat equation (5.20) to the second order in space and in time $\mathscr{O}(\Delta t^2 + \Delta x^2)$. Indeed, using the standard Taylor expansions we obtain

$$\mathscr{L}_j^n = \tau u_{tt} + u_t - \nu u_{xx}$$
$$- \frac{\nu}{12} \Delta x^2 u_{xxxx} + \Delta t^2 \left[\tfrac{1}{6} u_{ttt} + \tfrac{1}{12} \tau u_{tttt} \right] + \mathscr{O}(\Delta t^4 + \Delta x^4).$$

The stability of the scheme (5.23) was studied in [35] and the following stability condition was obtained:

$$\frac{\Delta t}{\Delta x} \leqslant \sqrt{\frac{\tau}{\nu}}.$$

By taking, for example, $\tau = \nu \Delta x$ we obtain the following stability condition

$$\Delta t \leqslant \Delta x^{\frac{3}{2}},$$

which is still weaker than the standard parabolic condition (5.7). However, it was reported in [34, 36] that stable computations (even in 3D) can be performed even with $\Delta t = \mathscr{O}(\Delta x)$. The authors explain informally this experimental observation by the fact that usual stability conditions are too 'pessimistic'.

Remark 5.3 The ad-hoc parameter τ can be chosen in other ways as well. One popular choice consists in taking

$$\tau = \frac{\Delta x}{c_s},$$

where c_s is the real physical information speed.

5.3.3.3 Error Estimate

It is legitimate to ask the question how far are solutions $u_h(x, t)$ to the hyperbolic equation (5.20) from the solutions $u_p(x, t)$ of the parabolic heat equation (5.1) (for the same initial condition). This question for the initial value problem was studied in [144] and we shall provide here only the obtained error estimate. Let us introduce the difference between two solutions:

$$\delta u\,(x,\,t) \overset{\text{def}}{:=} u_h\,(x,\,t) \,-\, u_p\,(x,\,t).$$

Then, the following estimate holds

$$|\,\delta u\,(x,\,t)\,| \;\leqslant\; \tau\,\mathscr{M}\left(1\,+\,\frac{2}{\sqrt{\pi}}\right)\left(8\sqrt{2}\,\tau\,+\,\frac{\sqrt[4]{2\pi^2}}{2}\,T\right),$$

where $T\,>\,0$ is the time horizon and

$$\mathscr{M} \overset{\text{def}}{:=} \sup_{\Omega_{\xi,\zeta}}\left|\frac{\partial^2 u_p}{\partial t^2}(\xi,\,\zeta)\right|,$$

and the domain $\Omega_{\xi,\zeta}$ is defined as

$$\Omega_{\xi,\zeta} \overset{\text{def}}{:=} \left\{(\xi,\,\zeta)\,:\,0\,\leqslant\,\zeta\,\leqslant\,t,\quad x\,-\,\frac{t-\zeta}{\sqrt{\tau}}\,\leqslant\,\xi\,\leqslant\,x\,+\,\frac{t-\zeta}{\sqrt{\tau}}\right\}.$$

5.4 Discussion

We saw above that only explicit schemes allow to have the finite speed of information propagation in the discretized version of the heat equation (5.1). The ease of parallelization along with excellent scaling properties of the codes obtained with explicit schemes constitute another important advantage to privilege explicit schemes over implicit ones [35]. However, explicit schemes for parabolic equations suffer from very stringent CFL-type conditions $\Delta t\,=\,\mathcal{O}(\Delta x^2)$ on the time step. That is why it is almost impossible to perform long time simulations (required in e.g. building physics applications) using fully explicit schemes such as (5.5). In order to keep explicit discretizations and overcome stringent CFL-type restrictions, a certain number of *hybrid* schemes were described. These hybrid schemes are based on different ideas. Some schemes rely on the information about the numerical solution on following time layers while keeping the overall scheme explicit using various tricks. Some hybrid schemes (e.g. Dufort–Frankel method described in Sect. 5.3.1 and Saulyev's scheme from Sect. 5.3.2) can be even unconditionally stable.

   Using the local error analysis we noticed that CFL-improved schemes gain the stability by introducing some weak hyperbolicity into the model (5.2). In particular, it is the case of the Dufort–Frankel scheme, while Saulyev method seems to be rather dispersive. This observation suggests that a new method can be derived by introducing this hyperbolicity in a controlled manner and to discretize the perturbed equation later. The method of hyperbolization deforms the equation operator to achieve desired properties[4] of the numerical solution. However, in all cases the gain in stability results in some loss of accuracy in representing the original continuous operator (5.1). The trade-off between the stability and accuracy has to be made by the end user. The second order accuracy is equivalent to the classical parabolic CFL-type condition. However, the user can choose to degrade intentionally the accuracy to relax the stability restriction up to hyperbolic-type conditions $\Delta t = \mathscr{O}(\Delta x)$ (originally found in [43]) and even beyond. It is not difficult to see that hybrid schemes presented in this study can be easily generalized to nonlinear cases with source terms in one and more spatial dimensions.

The main goal of this chapter was to communicate and attract community's attention to these improved discretizations, which can be used in modern building physics simulations where typical time scales are measured rather in months or even years. Our preference goes perhaps to the method of hyperbolization since it has been successfully applied (and validated) even to compressible Navier–Stokes and MHD equations [34]. Another advantage of hyperbolization technique is that it can be mathematically derived for gas dynamics from the kinetic theory of Boltzmann.

---

[4]By desired properties we mean the finite speed of information propagation along with less stringent CFL-type restrictions in explicit finite difference discretizations.

Chapter 6
Reduced Order Methods

6.1 Introduction

In building physics, as mentioned in Chaps. 2 and 3, numerical models used to predict heat and moisture transfer involve different characteristic time and lengths. Simulation of building behavior is generally analyzed on a time scale of 1 year (or more). However, the phenomena and particularly the boundary conditions evolve in seconds. The geometric configurations of the buildings require three-dimensional modeling of a facade of several meters. Furthermore, when dealing with heat and moisture, the nonlinear behaviors of the materials might be taken into consideration. Thus, the numerical model may require fine discretization of space and time domains to solve large algebraic systems. Model reduction be used to approach the solution of the problem in a reduced space dimension. This chapter aims at introducing model reduction methods to be applied to diffusion problems in applications of building physics.

6.1.1 Physical Problem and Large Original Model

The physical problem involves a heat conduction problem in a 1-dimensional domain Ω. The thermophysical properties of each region are supposed to be constant inside the entire domain. The contact between neighboring regions Ω_i and Ω_j, $\forall i \in \{1, \ldots, I\}$, $\forall j \in \{1, \ldots, I\}$, $i \neq j$ is assumed to be perfect. The initial temperature in the body is $u_0(x)$. The heat conduction problem can be written as follows:

$$\frac{\partial u}{\partial t} - v \frac{\partial^2 u}{\partial x^2} = 0 \qquad x \in \Omega, \, t > 0, \qquad (6.1)$$

© Springer Nature Switzerland AG 2019
N. Mendes et al., *Numerical Methods for Diffusion Phenomena in Building Physics*,
https://doi.org/10.1007/978-3-030-31574-0_6

where u is the field (as temperature or vapor pressure) and ν is the diffusivity of the domain. At the interface air–material, three types of boundary conditions can be defined: (i) the surface Γ_D is maintained at a temperature of u_D (Dirichlet condition), (ii) a heat flux q is imposed at surface Γ_N (Neumann condition) and (iii) a mixed condition is imposed at surface Γ_R (Robin condition). For clarity, all the content is devoted to the 1D heat transfer problem (6.1). Nevertheless, the extension to two and three dimensions is natural.

We assume that problem (6.1) can be solved using any classical numerical method, such as the conventional methods presented in Chaps. 3 and 4. The solution of this problem is denoted as $u(x, t)$, and the model used to obtain it is called the **Large Original Model** (LOM). The resulting discrete system of the LOM can be written as:

$$\mathbf{K}\,\mathbf{U}^{m+1} = \mathbf{L}\,\mathbf{U}^m + \mathbf{B}\,\mathbf{Q}, \tag{6.2}$$

where $\mathbf{U}$ is the discrete representation of field u, $\mathbf{K}$ and $\mathbf{A}$ are the matrices of the discrete interpolation of operators, $\mathbf{Q}$ is the vector containing the thermal inputs (boundary conditions) of the problem and matrix $\mathbf{B}$ links the discretization nodes to the thermal inputs $\mathbf{Q}$. We define the **order** M of the LOM as the dimension or degree of freedom of the solution U. In usual building physics applications, M can be of the order 10^2 up to 10^6.

6.1.2   Model Reduction Methods for Building Physics Application

Model reduction techniques, rather than a LOM, can be used to approximate the solution in lower-dimensional space. The solution obtained using a **Reduced Order Model (ROM)** has a reduced number of equations to solve. Note that by using the ROM, $\tilde{u}$ is obtained as the solution of problem (6.1) and U_r is its discrete representation.

Figure 6.1 illustrates two families of methods. The first is called a posteriori and requires a computed (or experimental) solution of the problem to build the ROM. After the so-called learning step, the ROM can be used to carry numerical prediction for other physical problems. This family comprises two interesting aspects. First, the ROM is created using solutions of the LOM for a short time interval and the ROM is then used for simulation on a longer time interval. Second, the ROM is created for a defined time interval and used for problems with the same time intervals but different boundary conditions or material properties.

The second family of reduction techniques is the a priori methods. These techniques do not need preliminary information on the studied problem. The ROM is an unknown a priori and is built directly.

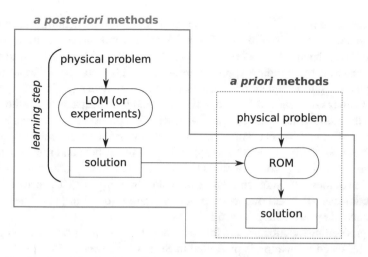

Fig. 6.1 Family of model reduction techniques

Different model reduction methods have been used for building physics applications. The first a posteriori technique was called the **Modal Identification Method** (MIM) and has been used in several linear and nonlinear inverse heat conduction problems. In [79, 196], a 2D and 3D heat conduction is treated. Furthermore, in [75–77], 2D and 3D forced convection problems were addressed. The basic principle is to identify a reduced basis of the model with observed (numerical) data. The ROM is built through an optimization problem. The details of this method will be presented in Sect. 6.3.

The **Proper Orthogonal Decomposition (POD)** is another a posteriori technique that has been widely used in different linear and nonlinear problems in many scientific fields. POD has been used to compute a ROM with Navier–Stokes equations in [57, 129] and with a building physics applications in [179]. In addition, this technique was applied in heat conduction problems of thermal bridges in [203]. Applications for heat and moisture transfer can be found in [152] (1D linear case) and in [14] (2D nonlinear case). Further references and details are given in Sect. 6.4.

In contrast, four a priori techniques can be referred to for building physics applications. A widely used approach is the **balanced truncation**, which is based on developing a specific reduced basis for the model different from the original physical basis of state variables. A selection of the dominant modes is then operated to obtain the ROM. This method is presented in Sect. 6.2, and has been used for thermal systems and a multizone model in [83, 114], respectively. In [69], a ROM was built for heat conduction in thermal bridges, and integrated into transient system (TRNSYS) tools for performing entire building energy simulations.

Next, the so called **Branch Eigenmodes Reduction** (BER) builds the ROM by changing the basis and solving a specific spectral problem, called the Branch Eigenmodes problem. This problem corresponds to the spectral problem associated with the advection diffusion operator, considering a Steklov boundary condition. Further details can be found in [145, 197] with an application for the heat transfer problem. Furthermore, a comparison of the BER and MIM methods is described in [195].

In addition, the use of an a priori version of the POD has been applied to a phase change material for building applications in [50, 51]. Here the ROM is built by solving the Lyapunov equation in a control theory framework.

The **Proper Generalised Decomposition** (PGD) is also an a priori method, originated in the radial space time separated representation proposed by Ladeveze in 1985 [120]. It is assumed to be a separated representation solution of each coordinate of the problem. Recently, the PGD was applied to nonlinear heat and moisture transfer problems in building porous materials [15] and in multizone problems [16] with an integration of these ROM for simulating the whole building [17]. The use of the PGD to solve inverse 3D heat conduction problems can be found in [18]. The basics of this method are presented in Sect. 6.5.

The proceedings section present four methods. The construction of a reduced basis using a balanced truncation is presented in Sect. 6.2. Sections 6.3 and 6.4 provides details of the a posteriori methods using modal identification and POD, respectively. Finally, the basics of the PGD are discussed in Sect. 6.5. It should be noted that the Spectral method can also be considered as model reduction approach, which is presented further in Chap. 8

6.2   Balanced Truncation

6.2.1   Formulation of the ROM

The model reduction can be operated on the discrete system of the problem. The main concept is to develop a specific (reduced) basis model that is different from the original physical basis of state variables. Therefore, the discrete representation of Eq. (6.2) is used to write the LOM in the state space representation as follows [20]:

$$\dot{U} = A\,U + B\,Q, \qquad\qquad (6.3a)$$

$$Y = HU, \qquad\qquad (6.3b)$$

where Y is the matrix of the interesting outputs of the problem (for instance, it can be the temperature and/or flux at one or several points) and matrix H operates the transformation from temperature U to the outputs of interest. Using the discrete representation in Eq. (6.3), we determine the eigenvalues of matrix A by solving the following spectral problem:

$$A\,P = \lambda\,P, \qquad\qquad (6.4)$$

where P is the eigenvector associated with the eigenvalue λ of matrix A. Furthermore, M is the total number of eigenvalues and Φ is the eigenvector basis composed of M eigenvectors such that:

$$\Phi = \begin{bmatrix} \mathbf{P}_1 & \cdots & \mathbf{P}_M \end{bmatrix}. \tag{6.5}$$

The construction of the ROM is based on the truncation of the eigenvector basis obtained using the Marshall method.

6.2.2 Marshall Truncation Method

To perform a truncation on the eigenvector basis Φ from Eq. (6.5), the Marshall approach proposes that the eigenvectors associated with the greatest eigenvalues be retained. All the eigenvalues are real negative values with the dimension of t^{-1}. The time constant is defined as follows:

$$\tau_i = \frac{-1}{\lambda_i}, \qquad \forall\, i \in \{1, \ldots, M\}.$$

The so-called fast and slow modes are then classified, with the slow mode verifying:

$$\tau_i > \eta,$$

where η is a parameter fixed by the user according to the time discretization of the numerical resolution. The Marshall truncation selects only the slow eigenmodes of the eigenbasis Φ. The contribution of the remaining slow eigenmodes is assumed to be negligible.

6.2.3 Building the ROM

After operating the truncation using the Marshall method, the new reduced basis Φ_r composed of the eigenvector states can be written as

$$\Phi_r = \begin{bmatrix} \mathbf{P}_1 & \cdots & \mathbf{P}_N \end{bmatrix}, \tag{6.6}$$

with $N \ll M$. Owing to its association with Φ_r, $\mathbf{C}$ is the diagonal matrix containing N truncated eigenvalues. The discrete representation $\mathbf{U}_r$ of the approximated field $\tilde{u}$ is computed as

$$\mathbf{U} \simeq \mathbf{U}_r = \Phi\,\mathbf{X}, \tag{6.7}$$

where $\mathbf{X}$ is the new state vector of dimension N. Based on Eqs. (6.2) and (6.7), the ROM is stated as:

$$\dot{\mathbf{X}} = \mathbf{C}\,\mathbf{X} + \mathbf{D}\,\mathbf{Q}, \tag{6.8a}$$

$$\mathbf{Y}_r = \mathbf{H}_r\,\mathbf{X}, \tag{6.8b}$$

where $\mathbf{C} = \mathrm{diag}\,(\lambda_i)$, $\forall i \in \{1, \ldots, N\}$ is the diagonal matrix containing N truncated eigenvalues of the spectral problem (6.4), $\mathbf{D} = \Phi_r^{-1}\,\mathbf{B}$ is the matrix realizing the links with the thermal inputs, $\mathbf{H}_r = \mathbf{H}\,\Phi_r$ is the matrix operating the transformation to the interesting outputs of the problem and $\mathbf{Y}_r$ represents the interesting outputs of the problem computed by the ROM. Note that matrix $\mathbf{C}$ is diagonal and has a low dimension ($N \ll M$). Thus, the solution of Eq. (6.8) is easy to compute.

6.2.4 Synthesis of the Algorithm

To build the ROM by using the aforementioned method, we must follow the strategy described as follows.

---

1 Write the LOM in a discrete representation according to Eq. (6.2) ;
2 Solve the spectral problem (6.4) associated to matrix A ;
3 Perform the truncation and define the reduced basis Φ from Eq. (6.6) ;
4 State the ROM Eq. (6.8): $\dot{\mathbf{X}} = \mathbf{C}\,\mathbf{X} + \mathbf{D}\,\mathbf{Q}$ and $\mathbf{Y} = \mathbf{H}_r\,\mathbf{X}$;

Algorithm 1: Strategy to build the balanced truncation ROM.

6.2.5 Application and Exercise

For this application, we consider the dimensionless problem of 1D heat transfer (6.1) for $x \in [0, 1]$ and $t \in [0, 10]$. Neumann boundary conditions are imposed:

$$\nu\,\frac{\partial u}{\partial x} = q_L(t) \qquad\qquad x = 0,\, t > 0,$$

$$-\nu\,\frac{\partial u}{\partial x} = q_R \qquad\qquad x = 1,\, t > 0.$$

The following values of the parameters are considered for numerical applications:

$$\nu = 0.01, \quad q_L(t) = 2 \cdot 10^{-2}\,\sin\!\left(\frac{\pi}{5}\,t\right), \quad q_R = -0.01, \quad u_0(x) = x^2.$$

Furthermore, the following discretization parameters can be considered: $\Delta x = 0.01$ and $\Delta t = 2.5 \cdot 10^{-3}$. The discrete time grid is denoted as t_m. In this problem, we are interested in computing the field at points $x_i = \{0,\, 0.2,\, 0.9\}$.

Questions:

1. Define matrices A, B, Q, and H according to the notation of Eq. (6.3) and compute the eigenvector basis.
2. Solve the eigenvalue problem and operate the truncation in the eigenvector basis using the Marshall method and $\eta = \frac{1}{50}$.
3.   a. Compute the results of the ROM and compare with the LOM.
    b. Compare orders N and M of the ROM and LOM, respectively,
    c. Study the influence of η.

Solution:

1. The LOM is written as:

$$\dot{U} = A\,U + B\,Q,$$
$$Y = H\,U.$$

Using finite differences, matrices A, B and Q are defined as:

$$A = \frac{\nu}{dx^2}\begin{bmatrix} -1 & 1 & 0 & \cdots & 0 \\ 1 & -2 & 1 & \ddots & \vdots \\ 0 & \ddots & \ddots & \ddots & 0 \\ \vdots & \ddots & 1 & -2 & 1 \\ 0 & \cdots & 0 & 1 & -1 \end{bmatrix}, \quad B = \begin{bmatrix} 1 & 0 \\ 0 & 0 \\ \vdots & \vdots \\ 0 & 0 \\ 0 & 1 \end{bmatrix}, \quad Q = \frac{1}{dx}\left[q_L(t_m)\ q_R\right]^T.$$

In addition, matrix H is defined as

$$H = 0_{(3,101)}, \quad H(1,1) = 1, \quad H(2,21) = 1, \quad H(3,91) = 1,$$

where $0_{(3,101)}$ denotes a matrix with 3 rows and 101 columns filled with zeros. Note that $B \in \mathbb{R}^{101 \times 2}$, $Q \in \mathbb{R}^{1 \times 2}$, $A \in \mathbb{R}^{101 \times 101}$ and $H \in \mathbb{R}^{3 \times 101}$.

2. The solution of the spectral problem gives $M = 101$ eigenvalues and eigenvectors. The time constant τ_i, $i \in \{1, \ldots, 101\}$ are computed for each eigenvalue. The truncation $\tau_i > \frac{1}{50}$ is performed and $N = 24$ eigenvalues are selected, as illustrated in Fig. 6.2. The time constant τ_i is represented for $i \in \{1, \ldots, 101\}$.

3. The ROM is written as:

$$\dot{X} = C\,X + D\,Q,$$
$$Y = H_r\,X.$$

The order of the ROM is $N = 24$, corresponding to the size of the state variable X; whereas the order of the LOM is $N = 101$. Matrix C is diagonal and contains the $N = 24$ eigenvalues selected through the truncation. The size of matrix H_r is 3×2. Figure 6.3 presents the results of the ROM and LOM for the interesting outputs. In addition, Fig. 6.4 shows the L_2 error for the total field u computed using the ROM

Fig. 6.2 Illustration of the
Marshall truncation:
selection of 24 eigenvalues

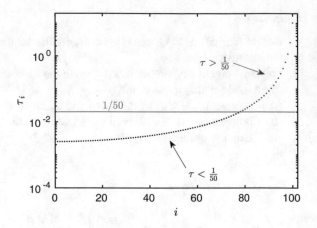

Fig. 6.3 Evolution of the
field u computed using the
LOM and ROM at the point
of interest

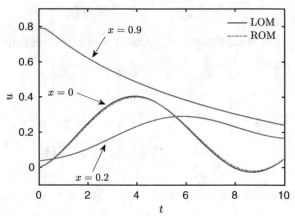

Fig. 6.4 Evolution of the
relative L_2 error between
solution u computed using
the LOM and ROM

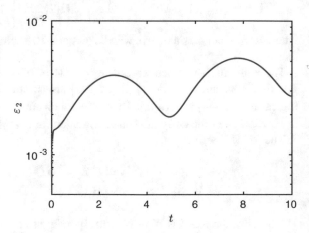

Fig. 6.5 Variation of the relative L_2 error between solution u computed using the LOM and ROM as function of the criterion η (**a**) and evolution of the order of the ROM as function of the criterion η

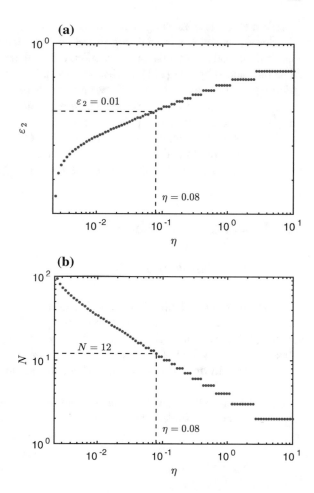

and LOM. As shown the error decreases with the time to reach an order of 10^{-2}. Figure 6.5 shows the influence of the criterion η on the error of the ROM solution as well as the order of the model. The L_2 error increases as the order of reduction of the ROM decreases. A ROM of order 12 seems to be a good compromise since it ensure an error of the order 10^{-2}.

6.2.6 Remarks on the Use of Balanced Truncation

The balanced truncation is based on the state space representation [20]. A change of basis is realized and a truncation is operated in the new basis. The solution of the problem is thus approximated in a lower space domain. The application shows that the size of the unknown solution is 24 and 101 for the ROM and LOM, respectively.

This method is interesting when focusing on outputs at particular points. The selection of modes is a crucial step in the construction of the reduced basis. The Marshall method is presented in this section. Other methods that can be used include those introduced by Litz [131] and Moore [143]. The modes are selected using the criteria of controllability, observability, and energy (with respect to the information sense) contained in the modes. Some interesting applications of this method in building physics are as follows. In [83], this model reduction method was applied to a multizone model for operating the control on HVAC[1] systems. In [68], a ROM was built for hollow bricks and the results obtained were compared with measurements. In [69], a ROM was developed for simulating 3D heat transfer in thermal bridges.

6.3   Modal Identification

6.3.1   Formulation of the ROM

The MIM aims at developing a ROM with a structure similar to that presented in Eq. (6.8) in Sect. 6.2. However, instead of computing matrices $\mathbf{C}$ and $\mathbf{H}_r$ by solving the spectral problem of matrix $\mathbf{A}$ (see Eq. (6.4)), they are sought through an identification process. An observation of the field $y_{\exp}$ is therefore required. This observation can be experimentally or numerically obtained and the method is therefore an a posteriori method.

N_q is the total number of independent inputs of the problem. An elementary ROM is built for each independent input q_k with $k \in \{1, \ldots, N_q\}$, and is thus written as

$$\dot{\mathbf{X}}_k = \mathbf{C}_k \mathbf{X}_k + \mathbf{1}_k q_k, \qquad \forall k \in \{1, \ldots, N_q\}, \qquad (6.9a)$$

$$\mathbf{Y}_r = \sum_{k=1}^{N_q} \mathbf{H}_{rk} \mathbf{X}_k, \qquad (6.9b)$$

where $\mathbf{X}_k$ is the new state vector of dimensions N_k, $\mathbf{C}_k$ is defined as a diagonal matrix, $\mathbf{Y}_r$ represents the interesting outputs of the problem with dimension $N_y \times N_k$, $\mathbf{H}_{rk}$ is the matrix operating the transformation between the state variable $X_{,k}$ and $\mathbf{Y}_r$ and $\mathbf{1}_k$ is a vector of size N, with all its components equal to 1. A total of N_q elementary ROMs compose the ROM, each with an order n. The order of the ROM is $N = \sum_{k=1}^{N_q} n_k$.

The approach to build an elementary ROM by using MIM is cast in a parameter estimation problem of matrices $\mathbf{C}_k$ and $\mathbf{H}_{rk}$ by minimizing the root mean quadratic discrepancy between the observation data $\mathbf{Y}_{\exp,k}$ and the elementary ROM outputs:

---

[1]Heating, Ventilation, and Air Conditioning.

$$\sigma(\mathbf{C}_k, \mathbf{H}_{rk}) = \sqrt{\frac{\left\| Y_r(\mathbf{C}_k, \mathbf{H}_{rk}) - Y_{\exp,k} \right\|_2^2}{N_q \times N_t}}. \tag{6.10}$$

The observation $\mathbf{Y}_{\exp,k}$ must be known for each thermal input q_k, while the other inputs are equal to zero. The minimization of $\sigma(\mathbf{C}_k, \mathbf{H}_{rk})$ depends on the order n of each elementary ROM. Therefore, it is rewritten as:

$$\sigma\left(n, \mathbf{C}_k, \mathbf{H}_{rk}\right) = \frac{\left\| Y_r(\mathbf{C}_k^{(n)}, \mathbf{H}_{rk}^{(n)}) - Y_{\exp,k} \right\|_2}{\sqrt{N_q \times N_t}}. \tag{6.11}$$

The estimation problem is considered solved when an order $n^\star$ verifies:

$$\sigma\left(n^\star, \mathbf{C}_k, \mathbf{H}_{rk}\right) \leq \varepsilon.$$

There are different possibilities for the choice of criterion ε. For experimental data, ε can be the standard deviation of the measurement errors. For a numerical application, ε is fixed as a function of the accuracy desired by the user.

6.3.2 Identification Process

For a fixed order n, matrices $\mathbf{C}_k^{(n)}$ and $\mathbf{H}_{rk}^{(n)}$ are estimated by minimizing the cost function as follows:

$$\mathscr{J}\left(n, \mathbf{C}_k, \mathbf{H}_{rk}\right) = \left\| Y_r\left(\mathbf{C}_k^{(n)}, \mathbf{H}_{rk}^{(n)}\right) - Y_{\exp,k} \right\|_2 \leq \eta. \tag{6.12}$$

For this purpose, optimization algorithms are used. Note that output $\mathbf{Y}$ is linear with respect to $\mathbf{H}_{rk}^{(n)}$ and nonlinear with respect to $\mathbf{C}_k^{(n)}$. Thus, the following two types of optimization methods are used:

1 An iterative method is used for estimating $\mathbf{C}_k^{(n)}$, and can be either deterministic (e.g., conjugate gradient [4, 101, 103] and Levenberg–Marquardt [176, 204]) or stochastic (e.g., Genetic algorithm [165] and Bayesian inference [107]).

2 The ordinary least squares method can be used for estimating $\mathbf{H}_{rk}^{(n)}$ [11, 151] verifying:

$$\mathbf{Y}_{\exp,k} = \mathbf{H}_{rk}^{(n)} \mathbf{X}_k,$$

which can be written as:

$$\mathbf{H}_{rk}^{(n)} = \mathbf{Y}_{\exp,k} \frac{\mathbf{X}_k^T}{(\mathbf{X}_k \mathbf{X}_k^T)^{-1}}. \tag{6.13}$$

6.3.3   Synthesis of the Algorithm

Algorithm 2 presents the steps to build a ROM based on MIM using iterations on thermal inputs q_k and on the order n of the ROM. The order n is increased until the mean quadratic discrepancy $\sigma(\mathbf{C}_k, \mathbf{H}_{rk})$ is minimized (steps 3–12). Matrices $\mathbf{C}_k^{(n)}$ and $\mathbf{H}_{rk}^{(n)}$ are estimated in steps 4–9. In addition, matrix $\mathbf{H}_{rk}^{(n)}$ is computed using the ordinary least squares method in step 6. Step 8 shows that matrix $\mathbf{C}_k^{(n)}$ must be updated. This step is specific to the iterative method used. An initial guess is required for the first component of matrix $\mathbf{C}_k^{(n)}$ (step 2). This initialization can be performed randomly. Note that the initialization of matrix $\mathbf{H}_{rk}^{(n)}$ is not required if only the initial condition of the field is zero.

---

1 **for** $k = 1 : N_q$ **do**
2   initial guess for $C_k(1, 1)$;
3   **while** $\sigma(n, \mathbf{C}_k, \mathbf{H}_{rk}) \geq \varepsilon$ **do**
4   **while** $\mathscr{J}(n, \mathbf{C}_k, \mathbf{H}_{rk}) \geq \eta$ **do**
5   Compute $\mathbf{X}_k$ using Eq. (6.9a) ;
6   Compute $\mathbf{H}_{rk}$ using Eq. (6.13) ;
7   Compute $\mathscr{J}(n, \mathbf{C}_k, \mathbf{H}_{rk})$ using Eq. (6.12) ;
8   Update $\mathbf{C}_k^{(n)}$;
9   **end**
10   Compute $\sigma(n, \mathbf{C}_k, \mathbf{H}_{rk})$ using Eq. (6.11) ;
11          $n = n + 1$;
12   **end**
13 **end**
14 State the ROM as defined in Eq. (6.9);

Algorithm 2: *Strategy to build the MIM ROM*

6.3.4   Application and Exercise

6.3.4.1   Exercise 1

For this application, we consider the dimensionless problem of 1D heat transfer (6.1) for $x \in [0, 1]$ and $t \in [0, 10]$. The boundary conditions are Neumann type: Neumann boundary conditions are imposed:

$$v \frac{\partial u}{\partial x} = q_L(t) \qquad\qquad x = 0, t > 0,$$

$$-v \frac{\partial u}{\partial x} = q_R \qquad\qquad x = 1, t > 0.$$

with the following parameter values for numerical applications:

$$v = 0.01, \quad q_L(t) = 2 \cdot 10^{-2} \sin\left(\frac{\pi}{5}t\right), \quad q_R = -0.01, \quad u_0(x) = 0.$$

A spatial and time discretization of $\Delta x = 0.01$ and $\Delta t = 2.5 \cdot 10^{-3}$ are chosen. In this problem, we focus on computing the temperature at points $x_i = \{0, 0.2, 0.9\}$.

Questions:

1. Compute the observation of the interesting outputs using a LOM for each thermal input.
2. a. Build the MIM ROM using $\sigma = 1 \cdot 10^{-3}$. In a MATLAB environment, function fminunc, based on a Quasi-Newton method, can be used for iterative optimization.
   b. Comment the order of the elementary ROM.
3. Compute the results of the ROM and compare with those of the LOM.
4. Using the ROM built previously, compute the solution of a problem by considering $q_L(t) = 2 \cdot 10^{-2} \left(1 - e^{-t}\right)$. Other numerical values are not changed. Compare the results thus obtained with those of the LOM.

Solution:

1. The interesting outputs $\mathbf{Y} = \left[u(x = 0, t) \quad u(x = 0.2, t) \quad u(x = 0.9, t)\right]^T$ are computed using central finite differences and an Euler backward scheme. The first and second observations, that is $u_{\exp, L}$ and $u_{\exp, R}$ are computed by considering $q_r = 0$ and $q_L = 0$, respectively. The results are given in Fig. 6.6.
2. Using the previously computed observation, matrices $\mathbf{C}$ and $\mathbf{H}_r$ are identified using the MIM method. In a Matlab environment, function fminunc is used as shown in the following code:

```
1    Qk{1,1} = q_L; uexp{1,1} = u_expL;
2    Qk{2,1} = q_R; uexp{2,1} = u_expR;
3
4    for k=1:2      % iteration on the thermal inputs
5        q      = Qk{k,1};
6        % loading the observation corresponding to the thermal inputs
7        Y_exp = uexp{k,1};
8        % initialization of the order of the elementary ROM
9        n=1;
10       while sigma(n,k) > eta % iteration on the order of the elementary ROM
11
12           % initialization of the first component of C
13           cn = rand(1,1);
14
15           % minimization of the cost function by Quasi-Newton optimization
16           [cn, J] = fminunc(@(cn)cost_function(cn,n,q,Y_exp),cn,options);
17
18           % incrementation
19           ci = [ci cn];
20           n=n+1;
21           % computation of the mean quadratic discrepancy
22           sigma(n,k) = sqrt(J/(size(Hr,1)*Nt));
23
24       end
25
```

```
26      % storage of the matrices
27      sto_C{k,1}   = C;
28      sto_Hr{k,1}  = Hr;
29   end
```

Two elementary ROMs are built for each inputs q_L and q_R. The evolution of the error σ with the order of each elementary ROM is given in Fig. 6.7. For the criterion $\varepsilon = 1 \cdot 10^{-3}$, the two elementary ROMs have an order $N_1 = 5$ and $N_2 = 4$, respectively. Therefore, the order of the MIM ROM is $N = 9$.

3. The MIM ROM results are computed for the given numerical values and compared with the LOM results as shown in Fig. 6.8. The L_2 error is given in Fig. 6.9.

4. Using the previously built MIM ROM, the solution of a new problem is computed considering

$$q_L(t) = 2 \cdot 10^{-2} \left(1 - e^{-t}\right).$$

Figures 6.10 and 6.11 present the LOM and ROM results, respectively. As shown, the change of parameter q_L does not affect the accuracy of the ROM for this case study.

6.3.4.2　Exercise 2

This exercise aims at illustrating the use of the MIM ROM for solving an inverse problem. Here, we consider the dimensionless problem of 1D heat transfer (6.1) for $x \in [0, 1], t \in [0, 10]$ and Neumann type boundary conditions:

$$\nu \frac{\partial u}{\partial x} = q_L(t) \qquad\qquad x = 0, t > 0,$$

$$-\nu \frac{\partial u}{\partial x} = q_R \qquad\qquad x = 1, t > 0,$$

with

$$q_L(t) = A \sin\left(\frac{\pi t}{5}\right), q_R = -0.01,$$

and $u_0 = 0$, $\nu = 0.01$ and the following discretisation parameters $\Delta x = 0.01$ and $\Delta t = 2.5 \cdot 10^{-3}$. The issue is to determine parameter A using a MIM ROM, given an observation Y_{obs} at the interesting point $x_i = 0$. Table 6.1 lists the values of Y_{obs} at different times.[2] To solve the inverse problem, we aim at determining $A^\star$ and minimizing the residual $\mathscr{R}$:

$$A^\star = \arg\min\left(\mathscr{R}(A)\right), \tag{6.14}$$

---

[2] The observation has been computed through the LOM using finite central differences and the backward Euler scheme. A random noise is added to the 41 selected observations.

Fig. 6.6 Observation at the interesting point x_i for each thermal inputs q_L (**a**) and q_R (**b**)

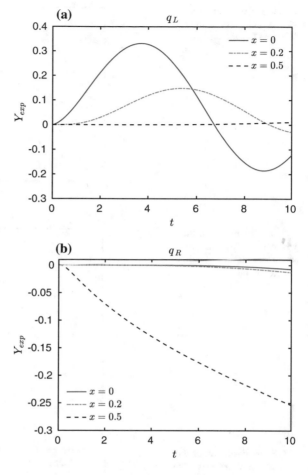

Fig. 6.7 Evolution of the mean quadratic discrepancy $\sigma(\mathbf{C}, \mathbf{H}_r)$ with the order of reduction of each elementary ROM (EROM)

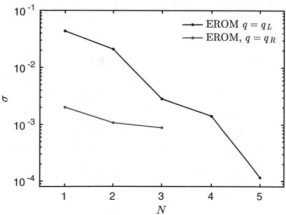

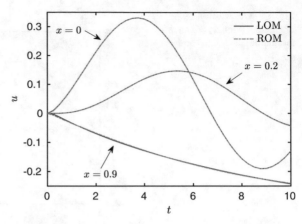

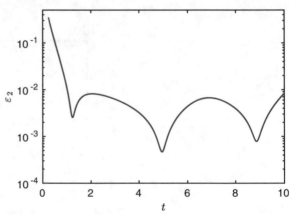

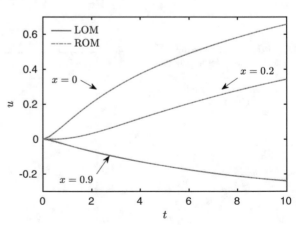

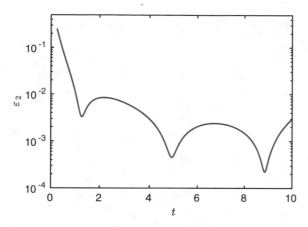

Fig. 6.11 Evolution of the relative L_2 error between solution u computed using the ROM and LOM for $q_L(t) = 2 \cdot 10^{-2} \left(1 - e^{-t}\right)$

Table 6.1 Observation values at $x = 0$ for different time intervals

Y_{obs}	t	Y_{obs}	t	Y_{obs}	t	Y_{obs}	t	Y_{obs}	t	Y_{obs}	t
0	0	0.48	2	0.71	4	0.26	6	−0.35	8	−0.3	10
0.04	0.25	0.54	2.25	0.7	4.25	0.18	6.25	−0.37	8.25		
0.06	0.5	0.6	2.5	0.66	4.5	0.09	6.5	−0.42	8.5		
0.11	0.75	0.65	2.75	0.62	4.75	−0.02	6.75	−0.43	8.75		
0.17	1	0.69	3	0.56	5	−0.12	7	−0.42	9		
0.27	1.25	0.72	3.25	0.5	5.25	−0.16	7.25	−0.41	9.25		
0.34	1.5	0.76	3.5	0.43	5.5	−0.26	7.5	−0.38	9.5		
0.4	1.75	0.73	3.75	0.35	5.75	−0.3	7.75	−0.35	9.75		

with the residual is defined as:

$$\mathscr{R}(A) = \left\| Y_{obs} - Y(A) \right\|_{L_2}, \tag{6.15}$$

where $Y(A)$ is computed using a MIM ROM.

Questions:

1. Build the MIM ROM for $A = 2 \cdot 10^{-2}$.
2. Compute the cost function $\mathscr{R}$ using Eq. (6.15) for 100 values of parameter A in the interval $\left[1 \cdot 10^{-2}, 1 \cdot 10^{-1}\right]$. Comment on the results.

Solution:

1. (see Exercise 1) The MIM ROM is built by considering $\varepsilon = 1 \cdot 10^{-3}$ and through Quasi-Newton optimization. Two elementary ROMs of order $N_1 = 5$ and $N_2 = 4$ are built for q_L and q_R, respectively. Matrices $\mathbf{C}_1$, $\mathbf{C}_2$, $\mathbf{H}_{r,1}$, and $\mathbf{H}_{r,2}$ are stored for solving the inverse problems.

Fig. 6.12 Evolution of the residual $\mathscr{R}$ with parameter A

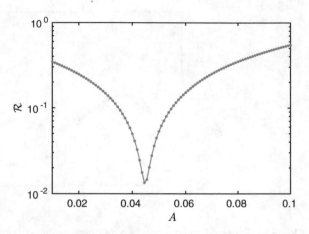

Fig. 6.13 Comparison of the observation and results of the MIM ROM for parameter $A^{\star}$

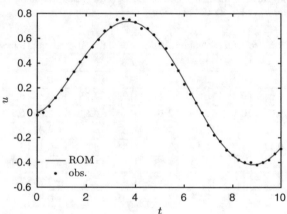

2. For 100 values of parameter A, the result Y is computed using the MIM ROM and matrices $\mathbf{C}_1, \mathbf{C}_2, \mathbf{H}_{r,1}$ and $\mathbf{H}_{r,2}$ that were computed previously. The variation of the residual $\mathscr{R}$ with parameter A is given in Fig. 6.12. The parameter A that minimize the cost function in the interval $\left[1 \cdot 10^{-2}, \ 1 \cdot 10^{-1}\right]$ is $A^{\star} = 4.5 \cdot 10^{-2}$. Figure 6.13 shows the observation and results of the ROM computed for $A^{\star}$.

6.3.5   Some Remarks on the Use of the MIM

We briefly presented the construction of a ROM using the MIM. It is cast in a parameter estimation problem of matrices composing the ROM. The estimation is performed using an optimization algorithm and by minimizing the cost function between the solution of the ROM and the observation field. An interesting aspect of this method is that the ROM is only built for the points of interest. Complete knowledge of the LOM is not required; only the results data (obtained from experiments or simulations) at the points of interest are needed. This feature is particularly interesting for solving inverse problems, as highlighted in Exercise 2 in Sect. 6.3.4.2. The ROM is built only for the point $x = 0$, and its solution is used for solving an inverse problem of the boundary heat flux. The studies of Girault, Petit, and Videcoq may be consulted for the interesting application of the method to heat conduction and/or convection problems [75–78, 148, 159, 195].

6.4 Proper Orthogonal Decomposition

6.4.1 Basics

The POD , also referenced as Karhunen–Loève decomposition, was proposed by Karhunen in 1946 [111] and Loève in 1955 [132]. This method aims at extracting the main information from a large set of data by projecting it into a smaller subspace. This method was developed and applied in a number of domains, such as image processing, oceanography, and weather prediction, and is one of the most widely used methods for reducing the order of dynamic models.

The objective of the POD is to search for a set of basis functions Φ, for which the projection of the field $u(x, t)$ best approximates the mean field [6]. To build the POD ROM, the method is divided into two steps inherent to the a posteriori characteristic. The first step uses the solution of a given problem obtained by the LOM (or experimental set up) on the discrete space and time domains. This step is generally called the excitation or learning step. The POD reduced basis is then built using the solution $u_{\exp}(x, t)$. In general, the field depends on the time and space coordinates; there are two possibilities of building the reduced basis: (1) the *classic* method in which the mean of field $u_{\exp}(x, t)$ is calculated over time, and (2) the *snapshots* approach, which uses a spatial mean of the field $u_{\exp}(x, t)$ at different times. Generally, in the framework of experimental measurements, data are treated using the *classic* method as illustrated in [63, 64]. For numerical experiments, as it is the case here, the *snapshots* procedure is recommended. Thus, the field $\tilde{u}(x, t)$ is approximated by:

$$\mathbf{U} \simeq \mathbf{U}_r = \Phi \mathbf{X}, \tag{6.16}$$

where Φ is the POD basis of order N and X is the unknown state of the ROM, depending only on the time variable. This method is detailed in further sections. However, readers may refer to [129] for specific aspects and [201] for more theoretical aspects.

6.4.2 Capturing the Main Information

Here, the task is to build an a posteriori basis Φ of the field. The basis Φ is composed of eigenvectors $\mathbf{P}$, verifying the spectral problem:

$$\mathbf{G}\,\mathbf{P} = \lambda\,\mathbf{P},\tag{6.17}$$

where $\mathbf{G}$ is the correlation matrix of matrix $\mathbf{Q}$ defined as:

$$\mathbf{G} = \mathbf{Q}\,\mathbf{Q}^T,\tag{6.18}$$

with $\mathbf{Q}$ being the *snapshots* matrix defined as:

$$Q_{ij} = U_{\exp}(x_i, t_j).\tag{6.19}$$

$U_{\exp}$ represents the observation of the problem used to build the a posteriori reduced basis obtained through experiments or numerically using the LOM.

6.4.3 Building the POD Model

After solving the spectral problem in Eq. (6.17), the POD ROM is built by performing a truncation in the basis Φ. Note that $\lambda_i, i \in \{1, \ldots, M\}$ in which the M eigenvalues are computed and sorted as

$$\lambda_1 \geqslant \cdots \geqslant \lambda_i \geqslant \cdots \geqslant \lambda_M.$$

Then cumulative energy $f(n)$ associated with the order n is then calculated as follows:

$$f(n) = \frac{\displaystyle\sum_{k=1}^{n} \lambda_k}{\displaystyle\sum_{i=1}^{M} \lambda_i}, \qquad\qquad n \in \{1, \ldots, M\}.\tag{6.20}$$

Furthermore, the truncation is operated at mode N, thus verifying

$$f(N) > \eta,$$

where η is a criterion selected by user. Thus, the POD basis of order N can be written as:

$$\Phi_r = \begin{bmatrix} P_1 & P_2 & \dots & P_N \end{bmatrix}.$$

Finally, the POD ROM is written as

$$\dot{X} = C X + D Q \tag{6.21a}$$

$$U_r = \Phi X, \tag{6.21b}$$

where $C = \Phi_r^{-1} A \Phi_r$, $D = \Phi_r^{-1} B$, and X is the vector of size N and the new unknown of the POD ROM. We also have X the new state vector of dimensions N, $C = \Phi_r^{-1} A \Phi_r$, $D = \Phi_r^{-1} B$.

6.4.4 Synthesis of the Algorithm

Owing to the POD strategy being *a posterior*, it is synthesized using the following algorithm, which is divided into 2 steps. The learning step aims at producing the results of fields $U_{\exp}$ (by using the LOM in our case). The POD ROM is then built by solving the spectral problem of the correlation matrix G (step 3).

---

1 *Learning step*: Compute $U_{\exp}$ using the LOM Eq. (6.2) ;
2 Calculate the correlation matrix G using Eq. (6.18) ;
3 Solve the spectral problem in Eq. (6.17) ;
4 State the ROM as defined in Eq. (6.21) ;

Algorithm 3: *Strategy to build the POD ROM.*

6.4.5 Application and Exercise

For this application, we consider the dimensionless problem of 1D heat transfer (6.1) for $x \in [0, 1]$ and $t \in [0, 10]$. Neumann boundary conditions are imposed:

$$\nu \frac{\partial u}{\partial x} = q_L(t) \qquad\qquad x = 0, t > 0,$$

$$-\nu \frac{\partial u}{\partial x} = q_R \qquad\qquad x = 1, t > 0,$$

and the numerical values:

$$\nu = 0.01, \qquad\qquad q_R = -0.01, \qquad\qquad u_0(x) = x^2.$$

The discretisation parameters are $\Delta x = 0.01$ and $\Delta t = 2.5 \cdot 10^{-3}$.

Questions:

1. a. *Simple learning step:* Compute the solution of the problem using the LOM for $A = 2 \cdot 10^{-2}$.
   b. Build the POD ROM choosing given $\eta = 0.99999$. Comment on the order of the ROM.
   c. Compute the results of the ROM and compare them with the LOM.
   d. Analyze the influence of the order N of the POD ROM on the error in the LOM.
3. a. Compare the results of the LOM and POD ROM previously computed (*Simple learning step* with $A = 2 \cdot 10^{-2}$), for $A = 3 \cdot 10^{-2}$.
   b. *Improved learning step:* Compute the solution of the problem using LOM, for $A = 1 \cdot 10^{-3}$ and $A = 4 \cdot 10^{-2}$. Then build the POD ROM.
   c. Compute the solution of the problem using the POD ROM built in the improved learning step for $A = 3 \cdot 10^{-2}$. Comment on the results.

Solution:

1. The problem is solved using a LOM based on finite differences and the backward Euler scheme, and the results are presented in Fig. 6.17a. These results are used to build the *snapshots* matrix. The spectral problem associated with the correlation matrix **G** is then solved. Figure 6.14 presents the cumulative energy $f(n)$ as a function of the order n of the ROM. An order of $N = 5$ is sufficient to reach $\eta = 0.99999$ of the total energy of the system.

The POD ROM is built using $\mathbf{C} \in \mathbb{R}^{5 \times 5}$ and $\Phi_r \in \mathbb{R}^{101 \times 5}$, and its solution is presented in Fig. 6.17b. Furthermore, Fig. 6.16 shows the error between both models. The results of the ROM and LOM are also compared for specific points in Fig. 6.15. The influence of the criterion η on the accuracy of the POD ROM is shown in Fig. 6.18. The error in the LOM decreases with order N of the ROM. A ROM of order 4 ensures an error lower than 10^{-2}.

2. The POD ROM built in the *simple learning step* ($A = 2 \cdot 10^{-2}$) is used to compute the solution of the problem given in Eq. (6.1) for 20 values of A in the interval

Fig. 6.14 Cumulative energy as a function of the order n of the POD ROM. The selected modes are highlighted in red color

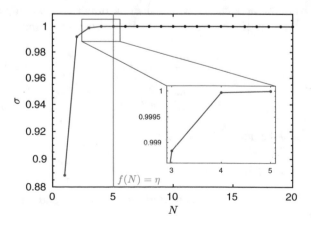

Fig. 6.15 Comparison of the LOM and ROM

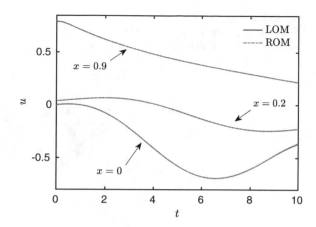

Fig. 6.16 Evolution of the relative L_2 error between the LOM and ROM

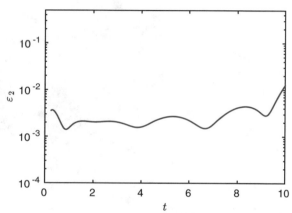

$[1 \cdot 10^{-3}, 4 \cdot 10^{-2}]$. The error in the LOM is calculated and illustrated in Fig. 6.19. As the learning step of the POD ROM was executed for $A = 2 \cdot 10^{-2}$, the POD ROM is observed to be inaccurate in computing the solution of Eq. (6.1) with other numerical values of A.

Here the task is to improve the POD basis. For this, the problem in Eq. (6.1) is solved for $A = 1 \cdot 10^{-3}$ and $A = 4 \cdot 10^{-2}$. The *snapshot* matrix is built by the concatenation of both results. The POD ROM is defined as described in Sect. 6.4.3. Then, it is used to compute the solution of Eq. (6.1) for 20 values of A in the interval $[1 \cdot 10^{-3}, 4 \cdot 10^{-2}]$. The error obtained by solving the LOM is given in Fig. 6.19. The accuracy of the POD ROM increases by improving the learning step.

Fig. 6.17 Solution of Eq.
(6.1) computed using the **a**
LOM and the **b** POD ROM

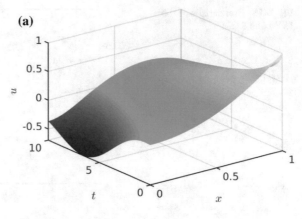

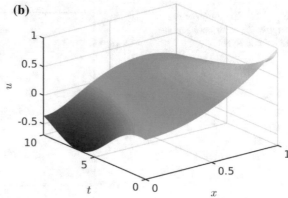

Fig. 6.18 Error of the ROM
as a function of the order N

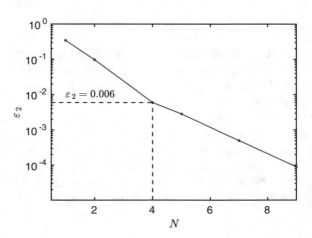

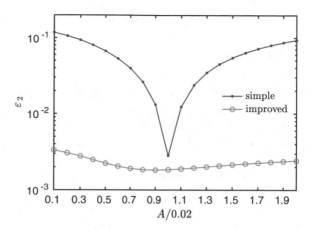

Fig. 6.19 Error of the POD ROM for computing the solution of Eq. (6.1) using one value of A (simple learning step) and two different numerical values of A (improved learning step)

6.4.6 Remarks on the Use of the POD

This section illustrates the use of the POD method for model reduction. The reduced basis is built using an observation of the field obtained numerically (*snapshots* approach). Applications have shown that a ROM of order 5 is sufficient for accurately computing the fields.

The *snapshots* approach of POD is particularly beneficial when the ROM can be used to solve a problem with different configurations than those used in the learning step. In the majority of the cases, the new problems consider slight variations in time interval, in material parameters, or in boundary conditions, for instance, in [6, 88, 154, 179]. However, building a robust POD basis capable of dealing with conditions different from those used for its construction (learning step) is important. This problem is illustrated in Exercise 1. Once it is solved, the POD ROM promises interesting computational gains.

In the framework of building applications, this method was first applied to computational fluid dynamics problems of turbulent flows by Lumley [133] in 1967. Other studies were then conducted on similar cases [19, 100], on the control of non-stationary flows [124, 202], on particle dispersion in a ventilated cavity [6], and on temperature distribution [179]. Some recent studies such as [123] combined the POD and PGD methods and applied them to Navier–Stokes equations. Moreover, other studies initiated the application of the POD method to heat diffusion [32, 153] and convection [154] problems.

6.5 Proper Generalized Decomposition

6.5.1 Basics

The PGD originated in the radial space-time separated representation proposed by Ladeveze in 1985 [120]. In 2006, the separated representations were extended to the multidimensional case by Chinesta et al. [7]. Interested readers may see Chinesta et al. [39, 41] for additional details on the method and [40] for an overview of its application.

The PGD basics assume that the solution of the problem $\tilde{u}(x, t)$ is approximated in the separated form as

$$U_r = \sum_{i=1}^{N} F_i(x) G_i(t).$$
(6.22)

Thus, the PGD approximation is a sum of N functional products involving functions (F_i, G_i). It is an a priori model reduction method and thus, a solution need not be computed before the problem to build the ROM. The approximation functions (F_i, G_i) are computed using an iterative procedure, as explained in further sections. Interested readers may refer to [41] for further details.

6.5.2 Iterative Solution

The calculation of problem (6.1) using the PGD method involves iteratively calculating modes (F_i, G_i) from $i = 1$ to $i = N$. The first mode (F_1, G_1) is initialised to verify the initial and boundary conditions.

At the enrichment step where $n < N$, we assume that a former approximation of $\tilde{u}(x, t)$ is known and the new couple $F_{n+1}(t) = R(x)$ and $G_{n+1}(x) = S(t)$ must be calculated according to:

$$U_r = \sum_{i=1}^{n} F_i(x) G_i(t) + R(x) S(t).$$
(6.23)

Equation (6.23) is introduced into Eq. (6.1). Owing to the separated representation of the solution $\tilde{u}$ for coordinates x and t, we have:

$$R \frac{dS}{dt} - \nu S \frac{d^2 R}{dx^2} = - \sum_{i=1}^{n} F_i \frac{dG_i}{dt} - G_i \frac{d^2 F_i}{dx^2} + \text{Res}^{n+1},$$
(6.24)

where Res^{n+1} is a residual because Eq. (6.24) is only an approximation of the solution.

6.5.3 Computing the Modes

At step n, we determine the new couple R and S by solving Eq. (6.24). To compute them, Eq. (6.24) is successively projected on R and S. Note that $(\bullet, \bullet)_t$ the scalar product associated to the time coordinate t and $(\bullet, \bullet)_x$ the scalar product associated to the space coordinate x. Equation (6.24) is projected on S, assuming $(\text{Res}^{n+1}, S)_t = 0$ to obtain:

$$\alpha_1 R + \beta_1 \frac{d^2 R}{dx^2} = \gamma_1, \qquad (6.25)$$

where

$$\alpha_1 = \left(S, \frac{dS}{dt}\right)_t$$
$$\beta_1 = v\left(S, S\right)_t$$
$$\gamma_1 = \sum_{i=1}^{n} -\left(S, \frac{dG_i}{dt}\right)_t F_i + v\left(S, G_i\right)_t \frac{d^2 F_i}{dx^2}.$$

Equation (6.24) is now projected on R, assuming $\left(\text{Res}^{n+1}, R\right)_x = 0$ to give:

$$\alpha_2 \frac{dS}{dt} + \beta_2 S = \gamma_2, \qquad (6.26)$$

where

$$\alpha_2 = \left(R, R\right)_x$$
$$\beta_2 = v\left(R, \frac{d^2 R}{dx^2}\right)_x$$
$$\gamma_2 = \sum_{i=1}^{n} -\left(R, F_i\right)_x \frac{dG_i}{dt} + v\left(R, \frac{d^2 F_i}{dx^2}\right)_x G_i.$$

After these projections, to solve equations an alternating direction fixed-point algorithm is used to solve Eqs. (6.25) and (6.26). The algorithm converges when the following stopping criterion is reached:

$$\text{Err}^q = \left|\left| (R\,S)^q - (R\,S)^{q-1} \right|\right| < \varepsilon, \qquad (6.27)$$

where q is the index of iteration of the fixed-point algorithm and ε is a parameter chosen by the user.

6.5.4 Convergence of Global Enrichment

After functions R and S are computed using the fixed-point algorithm, the PGD basis is enriched, displaying $F_{n+1} = R$ and $G_{n+1} = S$ as the new modes. The field of interest U_r can be written as

$$U_r = \sum_{i=1}^{n} F_i(x_i) G_i(t) + R(x) S(t) = \sum_{i=1}^{n+1} F_i(x) G_i(t).$$

The enrichment of the PGD solution stops when the norm of the residual Res^N is assumed to be negligible with respect to η a parameter chosen by the user

$$\left\| \mathrm{Res}^N \right\| = \left\| \sum_{i=1}^{N} \frac{\mathrm{d}G_i}{\mathrm{d}t} F_i - \nu G_i \frac{\mathrm{d}^2 F_i}{\mathrm{d}x^2} \right\| \leqslant \eta. \qquad (6.28)$$

6.5.5 Synthesis of the Algorithm

The strategy to build the PGD ROM is a priori and the approximation functions $\left(F_i, \ G_i \right)$ are computed using an iterative procedure as detailed in Algorithm 4.

---

1 Initialize F_1 and G_1 ;
2 while $\mathrm{Res}^N \geqslant \eta$ **do**
3   First guess for $\{R, \ S\}$;
4   **while** $\mathrm{Err}^q \geq \varepsilon$ **do**
5   Compute R using Eq. (6.25) ;
6   Compute S using Eq. (6.26) ;
7   Compute Err^q using Eq. (6.27) ;
8   **end**
9   Write $F_{n+1} = R, G_{n+1} = \mathscr{S}$;
10   Compute Res^N using Eq. (6.28) ;
11 end
12 State $U_r = \sum_{i=1}^{N+1} F_i G_i;$

Algorithm 4: *Strategy to build the PGD ROM.*

6.5.6 *Application and Exercise*

The problem from Eq. (6.1) is studied for $x \in [0, 1]$ and $t \in [0, 10]$. The boundary conditions are:

$$v \frac{\partial u}{\partial x} = 0.05 \left(\cos \left(\frac{\pi t}{5} \right) - 1 \right), \qquad x = 0, \ t > 0,$$

$$-v \frac{\partial u}{\partial x} = -0.09 \left(\cos \left(\frac{\pi t}{5} \right) - 1 \right), \qquad x = 1, \ t > 0,$$

and the following numerical values:

$$v = 0.01, \qquad\qquad u_0(x) = x^2.$$

and the discretisation parameters are $\Delta x = 0.01$ and $\Delta t = 2.5 \cdot 10^{-3}$.

Questions:

1. a. Compute the results of the ROM and compare them with those of the LOM.
   b. Analyze the influence of the number of modes N of the PGD ROM on the error in the LOM.

Solutions:

The problem is solved by considering the PGD method, central finite differences, and tolerances $\varepsilon = \eta = 1 \cdot 10^{-6}$, and the results are presented in Fig. 6.21a and b. In addition, the results of the LOM and PGD ROM are compared for particular points, as shown in Fig. 6.20, with the error between both models given in Fig. 6.22. The error with the LOM decreases with the number of ROM modes, as illustrated in Fig. 6.23. An order of 14 modes is sufficient to obtain a convenient accuracy of the order 10^{-2}.

Fig. 6.20 Comparison of the LOM and ROM

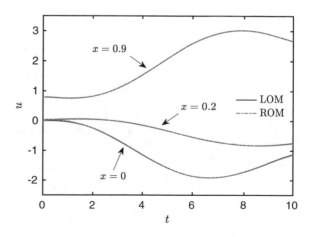

Fig. 6.21 Solution of the problem in Eq. (6.1) computed using the **a** LOM and the **b** POD ROM

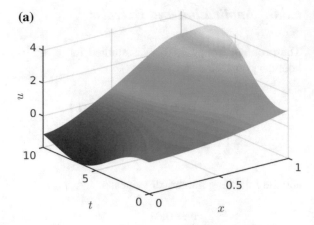

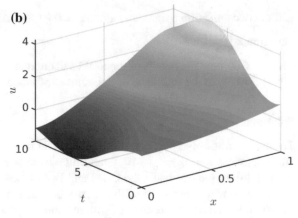

Fig. 6.22 Evolution of the relative L_2 error between the LOM and ROM

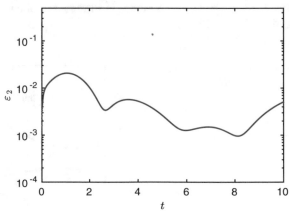

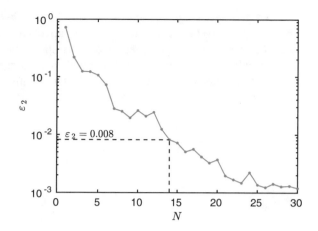

Fig. 6.23 Evolution of the relative L_2 error as a function of the number of modes of the PGD ROM

6.5.7 Remarks on the Use of the PGD

The PGD is an a priori model reduction method based on the search of a separated representation of the solution of the problem. The approximation functions are computed using an iterative procedure and an application is presented in Exercise 1.

Recently, the PGD was applied to non-linear heat and moisture transfer problems in building porous materials [15] and to multizone problems [16] with an integration of these ROMs for whole building simulation [17]. The use of the PGD in the solution of an inverse problem of 3D heat conduction can be found in [18].

6.6 Final Remarks

This chapter provided an introduction to model reduction methods for diffusion problems. In addition, balanced truncation, MIM, POD and PGD were presented. All these methods enable the determination of the solution of a problem in a reduced space dimension. The issue of reduction of the computational time is often arises when dealing with model reduction methods. However, this notion was not treated in this chapter mainly for the following reasons. First, the reduction of the computational time strongly depends on the algorithm writing and its optimization. It is important to comment and compare the number of operations accomplished by the algorithms. The highlight of computational benefits of reduced order models is especially observed for multidimensional models, which we believe are a trend in building physics in the upcoming years.

Some practical simple exercises are provided to help the readers in applying the methods as well as to illustrate the main advantages of the methods. Interested readers may refer to the given references to improve their knowledge, particularly, and particularly, to expand the method for coupled equations such as the HAM governing equations presented in Chap. 2. The treatment of nonlinearities of the material properties is also an important issue, as in any other method.

Chapter 7
Boundary Integral Approaches

In many engineering problems, including building modeling, the relevant information lies at the surface of a domain. In addition, only a few point-wise evaluations may be needed. Thus, classical numerical approaches would require a whole domain computation, which requires a considerable amount of information (and computation effort). The idea of using a boundary integral approach is to reduce the computation domain by using an analytic expression of the unknown field involving surface computation. In this chapter, two techniques are presented:

1. The Boundary Elements Method (BEM) or the Boundary Integral Equations Methods (BIEM) based on the finite-element approach. For an introductory course, the reader may refer to [24, 112]. This subject has been deeply investigated for several decades. A historical survey is presented in [33]. Its engineering applications can be found in [23, 134]; specifically, those for a transient diffusion process can be found in [49]. Specific techniques for integral equations are presented, e.g. in [52].
2. The Trefftz method (originally presented in [190]) has also been widely investigated. This approach belongs to the family of *meshless* methods, as the BEM, but uses collocation techniques. The reader may consult [97, 98, 168] for a presentation on the Trefftz method, and [128] or [116] for collocation techniques. A global survey of Trefftz methods is presented in [105, 127, 164] present its comparison with other boundary methods.

Hereafter, we focus on the 2D Poisson equation. Indeed, it is more convenient for presenting the methods. The addition of a particular variable-dependent source term transforms the Poisson equation into a Helmholtz equation (not treated in the exercises). It is easy to transform the semidiscretized heat equation into this Helmholtz equation.

Consider the 2D Poisson equation

$$-\nabla^2 u = f, \tag{7.1}$$

with mixed boundary conditions

© Springer Nature Switzerland AG 2019
N. Mendes et al., *Numerical Methods for Diffusion Phenomena in Building Physics*,
https://doi.org/10.1007/978-3-030-31574-0_7

$$u|_{\Gamma_D} = g(\boldsymbol{x}), \qquad \text{and} \qquad (\boldsymbol{\nabla}_n u)|_{\Gamma_N} = \Upsilon(\boldsymbol{x}), \qquad (7.2)$$

where $\boldsymbol{x} \in \Omega \subset \mathbb{R}^2$, and the boundary Γ of the domain Ω is divided into the following types of boundary conditions: Γ_D, which is a Dirichlet-type boundary condition, and Γ_N, which is a Neumann-type boundary condition. The unit normal exterior vector is $\boldsymbol{n}$.

7.1   Basic BIEM

The main steps of the BIEM comprise the formulation of the field value u by using boundary integral terms. This formulation is obtained using a double integration by parts and the *fundamental solution, or the Green function*, of the operator (here, the Laplacian). Such an evaluation of u is formulated for a set of N points at the boundary of the domain. This yields N relations that can be solved using the finite-element approach.

7.1.1   Domain and Boundary Integral Expressions

Assume a *function* $\psi_z(\boldsymbol{x})$ so that the Poisson equation can be projected on to it as follows:

$$-\int_{\Omega} \nabla^2 u(\boldsymbol{x}) \cdot \psi_z(\boldsymbol{x}) \, \mathrm{d}\boldsymbol{x} = \int_{\Omega} f(\boldsymbol{x}) \cdot \psi_z(\boldsymbol{x}) \, \mathrm{d}\boldsymbol{x} \, .$$

By using integration by parts and the divergence theorem, the left-hand side (LHS) becomes

$$-\int_{\Omega} \nabla^2 u(\boldsymbol{x}) \cdot \psi_z(\boldsymbol{x}) \, \mathrm{d}\boldsymbol{x} = -\int_{\Omega} \boldsymbol{\nabla} \cdot \left[\boldsymbol{\nabla} u(\boldsymbol{x}) \cdot \psi_z(\boldsymbol{x}) \right] \mathrm{d}\boldsymbol{x} + \int_{\Omega} \left[\boldsymbol{\nabla} u(\boldsymbol{x}) \cdot \boldsymbol{\nabla} \psi_z(\boldsymbol{x}) \right] \mathrm{d}\boldsymbol{x},$$

$$= -\int_{\Gamma} \left[\boldsymbol{\nabla}_n u(\boldsymbol{x}) \cdot \psi_z(\boldsymbol{x}) \right] \mathrm{d}\boldsymbol{x} + \int_{\Omega} \left[\boldsymbol{\nabla} u(\boldsymbol{x}) \cdot \boldsymbol{\nabla} \psi_z(\boldsymbol{x}) \right] \mathrm{d}\boldsymbol{x},$$

The second term on the right-hand side (RHS) can be recast by, once again, using integration by parts and the divergence theorem:

$$\int_{\Omega} \left[\boldsymbol{\nabla} u(\boldsymbol{x}) \cdot \boldsymbol{\nabla} \psi_z(\boldsymbol{x}) \right] \mathrm{d}\boldsymbol{x} = \int_{\Omega} \boldsymbol{\nabla} \cdot \left[u(\boldsymbol{x}) \cdot \boldsymbol{\nabla} \psi_z(\boldsymbol{x}) \right] \mathrm{d}\boldsymbol{x} - \int_{\Omega} \left[u(\boldsymbol{x}) \cdot \nabla^2 \psi_z(\boldsymbol{x}) \right] \mathrm{d}\boldsymbol{x},$$

$$= \int_{\Gamma} \left[u(\boldsymbol{x}) \cdot \boldsymbol{\nabla}_n \psi_z(\boldsymbol{x}) \right] \mathrm{d}\boldsymbol{x} - \int_{\Omega} \left[u(\boldsymbol{x}) \cdot \nabla^2 \psi_z(\boldsymbol{x}) \right] \mathrm{d}\boldsymbol{x}.$$

Thus, the LHS of the Poisson equations (7.1) is finally composed of three terms:

$$-\int_{\Omega} \nabla^2 u \cdot \psi_z \, dx = -\int_{\Gamma} \left[\nabla_n u \cdot \psi_z\right] dx + \int_{\Gamma} \left[u \cdot \nabla_n \psi_z\right] dx - \int_{\Omega} \left[u \cdot \nabla^2 \psi_z\right] dx .$$

Recall that the boundary $\Gamma = \Gamma_D \cup \Gamma_N$, where some information is known from the initial problem in Eqs. (7.1) and (7.2). Thus, by using these boundary conditions, the LHS of the Poisson equation reads as follows:

$$\text{LHS} = -\int_{\Gamma_D} \left[\nabla_n u \cdot \psi_z\right] dx - \int_{\Gamma_N} \left[\Upsilon \cdot \psi_z\right] dx + \int_{\Gamma_D} \left[g \cdot \nabla_n \psi_z\right] dx +$$
$$\int_{\Gamma_N} \left[u \cdot \nabla_n \psi_z\right] dx - \int_{\Omega} \left[u \cdot \nabla^2 \psi_z\right] dx .$$

Let us introduce the flux $q = \nabla_n u$; then the previous relation becomes

$$\text{LHS} = -\int_{\Gamma_D} \left[q \cdot \psi_z\right] dx + \int_{\Gamma_N} \left[u \cdot \nabla_n \psi_z\right] dx - \int_{\Gamma_N} \left[\Upsilon \cdot \psi_z\right] dx +$$
$$\int_{\Gamma_D} \left[g \cdot \nabla_n \psi_z\right] dx - \int_{\Omega} \left[u \cdot \nabla^2 \psi_z\right] dx ,$$

whereas the RHS of the Poisson equation is

$$\text{RHS} = \int_{\Omega} \left[f(x) \cdot \psi_z(x)\right] dx .$$

Hereafter, for the convenience of computation, assume that the source term is null, that is, $f \equiv 0$, so that RHS = 0. However, we provide some important remarks about the boundary integral method in the case of a nonzero source term. Thus, the double-integrated Poisson equation, projected on a given ψ_z, becomes

$$\int_{\Omega} \left[u \cdot \nabla^2 \psi_z\right] dx = -\int_{\Gamma_D} \left[q \cdot \psi_z\right] dx + \int_{\Gamma_N} \left[u \cdot \nabla_n \psi_z\right] dx$$
$$-\int_{\Gamma_N} \left[\Upsilon \cdot \psi_z\right] dx + \int_{\Gamma_D} \left[g \cdot \nabla_n \psi_z\right] dx ,$$
$$(7.3)$$

where the two first integral terms of the RHS contain the unknown $u(x)$ and its flux $q(x)$, for $x \in \Gamma$, whereas the two last integral terms represent the known contributions when the *function* $\psi_z(x)$ is explicit.

Remark 15 We have almost obtained a boundary integral formulation, as the RHS of the previous relation contains only boundary integral terms. This could have been reached because of the dropped source term.

7.1.2 Green Function and Boundary Integral Formulation

Let us focus on that *function* $\psi_z(x)$ that must be determined explicitly. Assume that ψ_z is the *function*, such that for any $z \in \mathbb{R}^2$,

$$-\mathbf{\nabla}^2\psi_z(x) = \delta(x, z),$$

where $\delta(x, z)$ is the Dirac function whose value is zero when $x \neq z$ and moves toward infinity when $x - z = 0$. ψ is called the *Green function* or the *fundamental solution* associated to the (negative) Laplace operator $-\mathbf{\nabla}^2$. For many other operators, a fundamental solution is known explicitly. Moreover, admittedly,[1] for any function $h(x)$, the fundamental solution has the property

$$-\int_{\mathbb{R}^2} h(x)\mathbf{\nabla}^2\psi_z(x)\,\mathrm{d}x = \int_{\mathbb{R}^2} h(x)\delta(x, z)\,\mathrm{d}x = h(z).$$

Thus, the LHS in Eq. (7.3) becomes

$$\int_{\Omega} \left[u(x) \cdot \mathbf{\nabla}^2\psi_z(x)\right]\mathrm{d}x = -\int_{\Omega} \left[u(x)\delta(x, z)\right]\mathrm{d}x = \begin{cases} -u(z) & \text{if } z \in \Omega \\ -C(z)u(z) & \text{if } z \in \Gamma \\ 0 & \text{otherwise} \end{cases}$$

The constant $C(z)$ depends on the local shape of the boundary in the neighborhood of point z. For a smooth boundary, $C(z) \equiv \dfrac{1}{2}$. By assembling the LHS and RHS as computed previously, for any $z \in \Gamma$, we get

$$\frac{1}{2}\,u(z) = \int_{\Gamma_D} \left[q(x) \cdot \psi_z(x)\right]\mathrm{d}x - \int_{\Gamma_N} \left[u(x) \cdot \mathbf{\nabla}_n\psi_z(x)\right]\mathrm{d}x$$
$$+ \int_{\Gamma_N} \left[\Upsilon(x) \cdot \psi_z(x)\right]\mathrm{d}x - \int_{\Gamma_D} \left[g(x) \cdot \mathbf{\nabla}_n\psi_z(x)\right]\mathrm{d}x.$$

That is, the final boundary integral formulation, for any $z \in \Gamma$, is as follows:

$$\frac{1}{2}\,u(z) + \int_{\Gamma_N} \left[u(x) \cdot \mathbf{\nabla}_n\psi_z(x)\right]\mathrm{d}x = \int_{\Gamma_D} \left[q(x) \cdot \psi_z(x)\right]\mathrm{d}x + \mathscr{I}(x, z),$$

with the known boundary integral term (it does not contain unknown variables u or q),

$$\mathscr{I}(z) = \int_{\Gamma_N} \left[\Upsilon(x) \cdot \psi_z(x)\right]\mathrm{d}x - \int_{\Gamma_D} \left[g(x) \cdot \mathbf{\nabla}_n\psi_z(x)\right]\mathrm{d}x.$$

---

[1] In order to be convinced, one has to be introduced to some basics of the Distribution theory. Indeed, the fundamental solution ψ must be viewed not as a classical function but as a distribution.

Note that

$$\mathscr{H}(z) = \int_{\Gamma_N} \left[u(x) \cdot \nabla_n \psi_z(x) \right] \mathrm{d}x \qquad \mathscr{F}(z) = \int_{\Gamma_D} \left[q(x) \cdot \psi_z(x) \right] \mathrm{d}x.$$

Thus, the boundary integral formulation reads as

$$\frac{1}{2} u(z) + \mathscr{H}(z) = \mathscr{F}(z) + \mathscr{I}(z), \qquad \text{for any } z \in \Gamma.$$

7.1.3 Numerical Formulation

Here we apply the FEM on the 1D domain Γ, corresponding to the (closed) boundary of a simply connex region $\Omega \subset \mathbb{R}^2$. As we are already familiar with the $\mathbb{P}1$ element, let us use it once again. Assume a partition of Γ in M segments of K_i, where $i = 1, \ldots, M$. Then assume that the degrees of freedom coincide with the M extremities of all K_i. In addition, we have $z_1, z_2, \ldots, z_M$ theses nodes.[2] Moreover, suppose that there are M_D nodes associated with the Dirichlet boundary condition, then $M_N = M - M_D$ nodes are associated to the Neumann boundary condition. For simplicity, we may assume that from $i = 1$ to M_D, the boundary nodes z_i correspond to a Dirichlet-type condition. Thus, following the boundary integral formulation obtained previously, we have M relations to evaluate:

$$\frac{1}{2} u(z_i) + \mathscr{H}(z_i) = \mathscr{F}(z_i) + \mathscr{I}(z_i), \qquad \text{for } i = 1, \ldots, M.$$

Recall that the $\mathbb{P}1$ basis test functions ϕ_i are defined as linear piecewise functions that take the value 1 on z_i, and 0 otherwise. Assume a decomposition of the approximate fields u_h and q_h by $\mathbb{P}1$ test functions:

$$u(z) \simeq u_h(z) = \sum_{j=1}^{M} \bar{u}_j \phi_j(z), \qquad q(z) \simeq q_h(z) = \sum_{j=1}^{M} \bar{q}_j \phi_j(z).$$

The weights $\bar{u}_j$ and $\bar{q}_j$ correspond to the unknowns to be computed. As a result of using the boundary conditions in Eq. (7.2), some of the unknowns are already computed.

$$\bar{u}_j = g(z_i) \qquad \text{for } z_i \in \Gamma_D \qquad \text{and} \qquad \bar{q}_j = \Upsilon(z_i) \qquad \text{for } z_i \in \Gamma_N.$$

Thus, only the set of M nodes $\{\bar{q}_1, \ldots, \bar{q}_{M_D}, \bar{u}_{M_D+1}, \ldots, \bar{u}_M\}$ must be computed.

---

[2] Recall that z belongs to the real plan; thus, it has two coordinates $z = {}^{\top}(z_1, z_2)$.

- For z_i, $i = 1, \ldots, M_D$,

$$\mathcal{F}(z_i) = \int_{\Gamma_D} \left[q(x) \cdot \psi_{z_i}(x) \right] dx = \sum_{K_D} \int_{K_D} \left[\sum_{j=1}^{M} \bar{q}_j \phi_j(x) \cdot \psi_{z_i}(x) \right] dx$$

$$= \sum_{j=1}^{M} \bar{q}_j \left(\sum_{K_D} \int_{K_D} \left[\phi_j(x) \cdot \psi_{z_i}(x) \right] dx \right),$$

where $K_D = \{ K_i, i = 1, \ldots, M \mid K_i \cap \Gamma_D \neq \varnothing \}$.

As the support of any test function ϕ_i is $K_{i-1} \cup K_i$, the computation of $\mathcal{F}(z_i) \equiv \mathcal{F}_i$ involves all the unknown weights in Γ_D and two additional nodes induced by K_1 and K_{M_D} connected to $\bar{q}_M = \Upsilon(z_M) = \Upsilon_M$ and $\bar{q}_{M_D+1} = \Upsilon(z_{M_D+1} = \Upsilon_{M_D+1}$, respectively.[3] Then, for z_i, $i = 1, \ldots, M_D$, we obtain

$$\mathcal{F}_i = \sum_{j=1}^{M_D} \bar{q}_j \left(\sum_{K_D} \int_{K_D} \left[\phi_j(x) \cdot \psi_{z_i}(x) \right] dx \right) =$$

$$\bar{q}_1 \left(\sum_{K_D} \int_{K_D} [\phi_1 \cdot \psi_i] \, dx \right) + \cdots + \bar{q}_i \left(\sum_{K_D} \int_{K_D} [\phi_j \cdot \psi_i] \, dx \right) + \cdots \bar{q}_{M_D} \left(\sum_{K_D} \int_{K_D} [\phi_{M_D} \cdot \psi_i] \, dx \right)$$

$$= \bar{q}_1 \left[F_{i,1} \right] + \cdots + \bar{q}_j \left[F_{i,j} \right] + \cdots + \bar{q}_{M_D} \left[F_{i,M_D} \right].$$

where the notation $\psi_{z_i}(x) \equiv \psi_i$ is used.

The evaluation of each $F_{i,j}$ can be implemented using any relevant quadrature formula. Finally, for $i = 1, \ldots, M_D$, one has M_D equations involves exactly M_D unknowns of $\bar{q}_i$. These M_D equations can be recast under a matrix form as follows:

$$\mathbf{Fq} + \mathbf{p} = \begin{pmatrix} F_{1,1} & \cdots & F_{i,M_D} \\ \cdots & F_{i,j-1} \; F_{i,j} \; F_{i,j+1} & \cdots \\ F_{1,M_D} & \cdots & F_{M_D,M_D} \end{pmatrix} \begin{pmatrix} \bar{q}_1 \\ \vdots \\ \bar{q}_j \\ \vdots \\ \bar{q}_{M_D} \end{pmatrix} + \begin{pmatrix} -F_{1,M} \Upsilon_M \\ 0 \\ \vdots \\ 0 \\ -F_{M_D,M_D+1} \Upsilon_{M_D+1} \end{pmatrix}$$

- A relatively analogous computation for the $M_N = M - M_D$ Neumann nodes z_i, $i = M_D + 1, \ldots, M$, yields an equivalent full matrix system:

$$\mathcal{H}(z_i) = \int_{\Gamma_N} \left[u(x) \cdot \nabla_n \psi_z(x) \right] dx$$

$$= \sum_{K_N} \int_{K_N} \left[\sum_{j=1}^{M} \bar{u}_j \phi_j(x) \cdot \nabla_n \psi_{z_i}(x) \right] dx$$

---

[3]This occurs depending on the selection of nodes with respect to the boundary conditions.

$$= \sum_{j=1}^{M} \bar{u}_j \left(\sum_{K_N} \int_{K_N} \left[\phi_j(x) \cdot \nabla_n \psi_{z_i}(x) \right] dx \right),$$

where $K_N = \{K_i, i = 1, \ldots, M \mid K_i \cap \Gamma_N \neq \varnothing\}$. Again, as the support of any test function ϕ_i is $K_{i-1} \cup K_i$, the computation of $\mathcal{H}(z_i) \equiv \mathcal{H}_i$ involves all the unknown weights in Γ_N and two additional nodes induced by K_{M_D+1} and K_M connected to $\bar{u}_M = g(z_M) = g_M$ and $\bar{u}_1 = g(z_1) = g_1$, respectively. Thus, for $i = M_D + 1, \ldots, M$,

$$\mathcal{H}_i = \sum_{j=M_D+1}^{M} \bar{u}_j \left(\sum_{K_N} \int_{K_N} \left[\phi_j(x) \cdot \nabla_n \psi_{z_i}(x) \right] dx \right)$$

$$= \bar{u}_{M_D+1} \left(\sum_{K_N} \int_{K_N} \left[\phi_{M_D+1} \cdot \nabla_n \psi_i \right] dx \right) + \cdots + \bar{u}_M \left(\sum_{K_D} \int_{K_N} \left[\phi_M \cdot \nabla_n \psi_i \right] dx \right)$$

$$= \bar{u}_{M_D+1} \left[H_{i,M_D+1} \right] + \cdots + \bar{u}_i \left[H_{i,j} \right] + \cdots + \bar{u}_M \left[H_{i,M} \right].$$

The evaluation of each $H_{i,j}$ can be implemented using any relevant quadrature formula. Finally, for $i = M_D + 1, \ldots, M$, the $M_N = M - M_D$ equations include exactly M_N unknowns of $\bar{u}_i$. These M_N equations can be recast under a matrix form:

$$\mathbf{Hu} + \mathbf{r} = \begin{pmatrix} H_{M_D+1,M_D+1} & \cdots & H_{M_D+1,M} \\ \cdots & H_{i,j-1} \ H_{i,j} \ H_{i,j+1} & \cdots \\ H_{M_D+1,M} & \cdots & H_{M,M} \end{pmatrix} \begin{pmatrix} \bar{u}_{M_D+1} \\ \vdots \\ \bar{u}_j \\ \vdots \\ \bar{u}_M \end{pmatrix} + \begin{pmatrix} -H_{1,M_D} g_M \\ 0 \\ \vdots \\ 0 \\ -H_{M,1} g_1 \end{pmatrix}$$

Next, all the equations must be assembled to obtain the final matrix system. An important issue is that the matrix involved in the BIEM is complete. From a computational point of view, a reliable procedure is required for solving it.

7.2 Trefftz Method

The idea here is to suppose a finite formal expansion of the solution by using a functional basis called T-complete function:

$$u(x, t) \simeq \tilde{u}(x, t) \simeq \sum_{i=1}^{N} a_i u_i^\star(x, t) = a \cdot u^\star(x, t).$$

The weight coefficients $a = \{a_i\}_{i=1,\ldots,N}$ are the numerical unknowns that must be determined, whereas the *T-complete functions* are well identified, depending on the differential operator. Such *T-expansion* of the solution is to exactly verify the differential equation on the domain Ω but not the boundary operators. Thus, the weight coefficients a chosen are those that meet most of the boundary conditions of the original problem.

In the following introductory example, we consider the homogeneous Poisson equation, or the Laplace equation:

$$\nabla^2 u(x) = 0, \qquad\qquad \text{for} \quad x \in \Omega,$$

with the following boundary conditions:

$$u = \bar{u} \quad \text{on Dirichlet boundary } \Gamma_D, \qquad \frac{\partial u}{\partial n} = \bar{q} \quad \text{on Neumann boundary } \Gamma_N.$$
(7.4)

7.2.1   Trefftz Indirect Method

This method was originally proposed by Trefftz [190].

T-complete functions and residual definition. Assume a formal expansion of the approximated solution:

$$u(x) \simeq \tilde{u}(x) \simeq \sum_{i=1}^{N} a_i\, u_i^\star(x) = a \cdot u^\star(x),$$

where the T-functions evaluated at a given point $x = \rho e^{i\theta}$ are

$$u^\star = \left\{ 1, \ldots, \left(\rho e^{i\theta}\right)^k, \ldots, \left(\rho e^{i\theta}\right)^n \right\}.$$

Exercise 12 *Show that $\tilde{u}$ is a solution of the Laplace operator, regardless of the values of the weight coefficients a.*

The normal approximated flux is simply

$$\tilde{q}(x) = \sum_{i=1}^{N} a_i\, q_i^\star(x) = a \cdot q^\star(x),$$
(7.5)

where $q^\star$ is naturally the normal derivative of $u^\star$. Regardless the values of the weight coefficients, $\tilde{u}$ verifies the Laplace equation at every point in the entire domain Ω. However, $\tilde{u}$ (and $\tilde{q}$) may not meet the boundary conditions in Eq. (7.4) exactly; here the values of a are crucial. To choose these values *optimally*, the *residuals* R_D and R_N, associated with each boundary type, are defined as follows:

- For the Dirichlet boundary condition on Γ_D,

$$R_D \equiv \tilde{u} - \bar{u} = a \cdot u^\star(P) - \bar{u}(P), \qquad P \in \Gamma_D.$$

- For the Neumann boundary condition on Γ_N,

$$R_N \equiv \tilde{q} - \bar{q} = a \cdot q^\star(P) - \bar{q}(P), \qquad P \in \Gamma_N.$$

Now, assume a discretization of the boundary $\Gamma = \Gamma_D \cup \Gamma_N$ composed of M_D and $M_N = M - M_D$ points associated to the Dirichlet (Γ_D) and Neumann (Γ_N) boundaries, respectively. Thus, a set of N coefficients $a = \{a_i\}$ that minimize the $M = M_D + M_N$ residual functions must be determined.

$$R_D(P_j) \equiv a \cdot u^\star(P_j) - \bar{u}(P_j)$$

$$\equiv \sum_{i=1}^{N} a_i\, u_i^\star(P_j) - \bar{u}(P_j), \qquad \forall P_j, j = 1, \ldots, M_D \in \Gamma_D \qquad (7.6)$$

$$R_N(P_k) \equiv a \cdot q^\star(P_k) - \bar{q}(P_k)$$

$$\equiv \sum_{i=1}^{N} a_i\, q_i^\star(P_k) - \bar{q}(P_k), \qquad \forall P_k, k = 1, \ldots, M_N \in \Gamma_N. \qquad (7.7)$$

Matrix system formulation   There are many procedures available for determining the weight coefficient a. However, different algorithms usually do not have the same numerical properties (such as robustness, easy implementation, low computational cost, and accuracy). Let us cite two such methods:

- *Collocation method*. In this method, the principle is to force the residuals to be zero at every point of collocation P_i. Thus, the corresponding matrix system reads as

$$
\begin{pmatrix}
u_1^\star(P_1) & \cdots & u_i^\star(P_j) & \cdots & u_N^\star(P_1) \\
\vdots & & \vdots & & \vdots \\
u_1^\star(P_{M_D}) & \cdots & u_i^\star(P_{M_D}) & \cdots & u_N^\star(P_{M_D}) \\
q_1^\star(P_{M_D+1}) & \cdots & q_i^\star(P_{M_D+1}) & \cdots & q_N^\star(P_{M_D+1}) \\
\vdots & & \vdots & & \vdots \\
q_1^\star(P_M) & \cdots & q_i^\star(P_M) & \cdots & q_N^\star(P_M)
\end{pmatrix}
\begin{pmatrix}
a_1 \\
\vdots \\
\\
a_i \\
\\
a_N
\end{pmatrix}
=
\begin{pmatrix}
\bar{u}_1 \\
\vdots \\
\bar{u}_{M_D} \\
\bar{q}_{M_D+1} \\
\vdots \\
\bar{q}_M
\end{pmatrix}
$$

This approach has the great advantage of being realized very easily by using libraries included in many computational languages or scientific software. A special case exists for which $N = M$ yields a unique solution (as soon as the matrix remains invertible). It should always be possible to move from the undetermined case ($M < N$) to the overdetermined case ($M > N$) by selecting more collocation points. In this configuration, a *singular value decomposition* is relevant.

- Galerkin method: The principle of this method is to *project* the residuals R_D and R_N on *well-chosen weight functions* ω_D and ω_N, and force the projection to be null. In the present case, a relevant choice for ω_D and ω_N is the exact approximated solutions $\tilde{q}$ and $-\tilde{u}$, respectively. Thus, from

$$\int_{\Gamma_D} \left[\tilde{q} \, R_D \right] \mathrm{d}x - \int_{\Gamma_N} \left[\tilde{u} \, R_N \right] \mathrm{d}x = 0 \,,$$

the scalar equation is obtained as

$$
\begin{aligned}
0 &= \int_{\Gamma_D} \left[\tilde{q} \, (a \cdot u^\star - \bar{u}) \, \mathrm{d}x \right] - \int_{\Gamma_N} \left[\tilde{u} \, (a \cdot q^\star - \bar{q}) \, \mathrm{d}x \right] \\
&= \int_{\Gamma_D} \left[(a \cdot q^\star) \, (a \cdot u^\star - \bar{u}) \, \mathrm{d}x \right] - \int_{\Gamma_N} \left[(a \cdot u^\star) \, (a \cdot q^\star - \bar{q}) \, \mathrm{d}x \right] \,.
\end{aligned}
$$

A sufficient condition for this relation to hold is given by N relations obtained formally by dropping the left-factorized vector a:

$$0 = \int_{\Gamma_D} \left[^\top(q^\star) \, (u^\star \cdot a - \bar{u}) \, \mathrm{d}x \right] - \int_{\Gamma_N} \left[^\top(u^\star) \, (q^\star \cdot a - \bar{q}) \, \mathrm{d}x \right] \,,$$

where $^\top(.)$ denotes the transposed vector.

This system of equations can be recast into a matrix formulation by *factorizing* with another vector a and removing the boundary terms from the RHS:

$$\left[\int_{\Gamma_D} \left(^\top(q^\star) \, u^\star \right) \mathrm{d}x - \int_{\Gamma_N} \left(^\top(u^\star) \, q^\star \right) \mathrm{d}x \right] a = \int_{\Gamma_D} \left[^\top(q^\star) \bar{u} \right] \mathrm{d}x - \int_{\Gamma_N} \left[^\top(u^\star) \bar{q} \right] \mathrm{d}x \,.$$

The products $^\top(q^\star) \, u^\star$ and $^\top(u^\star) \, q^\star$ produce two matrices. By defining

$$K_{ij} = \int_{\Gamma_D} \left(q_i^\star u_j^\star \right) \mathrm{d}x - \int_{\Gamma_N} \left(u_i^\star q_j^\star \right) \quad f_i \, \mathrm{d}x = \int_{\Gamma_D} \left[q_i^\star \bar{u} \right] - \int_{\Gamma_N} \left[u_i^\star \bar{q} \right] \mathrm{d}x \,,$$

each of the N lines of the matrix system reads as follows, for $i = 1, \ldots, N$:

$$K_{ij} \, a_j = f_i$$

This method for determining the weight coefficients $a = \{a_i\}$ is accurate and robust.

7.2.2 Method of Fundamental Solutions

This method is sometimes called as the *modified Trefftz method*, and it belongs to the family of indirect methods. In this method, the idea is to consider the fundamental solutions of the operator (instead of the T-functions):

$$\tilde{u}(x) = a \cdot u^\star(x) \qquad \text{with} \qquad u^\star \equiv \text{fundamental solution.}$$

For the working example given by the Laplace equation for $x \in \Omega$, with boundary Γ, recall that the fundamental solution for the Laplace operator is given by the relation for any two points x_1 and x_2:

$$\Psi(x_1, x_2) = \frac{1}{2\pi} \ln \left(\frac{1}{d(x_1, x_2)} \right),$$

where $d(x_1, x_2) = \sqrt{\sum_j \left((x_1)_j - (x_2)_j \right)^2}$ represents the Euclidian distance between x_1 and x_2. This fundamental solution becomes singular as $d(x_1, x_2) \to 0$. In particular, to avoid singularities of the fundamental solution evaluated for any $P \in \Gamma$, a virtual strictly surrounding domain $\mathsf{D} \supset \Gamma$, with boundary $\mathscr{D}$, must be considered. Thus, for any points $Q \in \mathscr{D}$ and $P \in \Gamma$, $\Psi(P, Q)$ cannot be singular. Fix N points $Q_i \in \mathscr{D}$, and define the modified Trefftz expansion evaluated for all the points $P \in \Gamma$ as follows:

$$\tilde{u}(P) = \sum_{i=1}^{N} a_i u_i^\star(P) = \sum_{i=1}^{N} a_i \Psi(P, Q_i) =$$

$$= a_1 \frac{1}{2\pi} \ln \left(\frac{1}{d(P, Q_1)} \right) + \cdots + a_j \frac{1}{2\pi} \ln \left(\frac{1}{d(P, Q_j)} \right) + \cdots + a_N \frac{1}{2\pi} \ln \left(\frac{1}{d(P, Q_N)} \right).$$

By using the expression of the fundamental solution, the flux term $\tilde{q}$ defined in Eq. (7.5) can also be explicitly computed. Thus, the process is similar to that for the indirect Trefftz method described earlier.

- For the Dirichlet boundary condition on Γ_D, the Dirichlet residual is defined as

$$R_D \equiv \tilde{u} - \bar{u} = a \cdot u^\star(P) - \bar{u}(P), \qquad P \in \Gamma_D.$$

- For the Neumann boundary condition on Γ_N, the Neumann residual is defined as

$$R_N \equiv \tilde{q} - \bar{q} = a \cdot q^\star(P) - \bar{q}(P), \qquad P \in \Gamma_N.$$

In addition, the matrix system associated to the null residuals condition for M boundary points P_j must be solved.

7.2.3 Trefftz Direct Method

This approach seems to be similar to the BIEM formulation. Contrary to the indirect methods, in which the weight vector $a = \{a_i\}$ occurring in the Trefftz expansion must be computed by minimizing residuals at the domain boundary Γ, the Trefftz direct method involves interpolating functions $\Phi = \{\phi_i\}$ to expand u at the boundary:

$$u(x) \simeq \tilde{u}(x) = \sum_{i=1}^{M} \phi_i(x) \, u_i = \Phi(x) \cdot \mathbf{u} \qquad \text{for} \quad x \in \Gamma.$$

Therefore, the normal flux vector can be expanded using the same interpolating function $\Phi = \{\phi_i\}$:

$$q(x) \simeq \tilde{q}(x) = \sum_{i=1}^{M} \phi_i(x) \, q_i = \Phi(x) \cdot \mathbf{q} \qquad \text{for} \quad x \in \Gamma.$$

By including the Dirichlet and Neumann boundary conditions, we get

$$\bar{u}(x_D) = \Phi(x_D) \cdot \mathbf{u} \qquad \text{for} \quad x_D \in \Gamma_D$$
$$\Upsilon(x_N) = \Phi(x_N) \cdot \mathbf{q} \qquad \text{for} \quad x_N \in \Gamma_N.$$

The unknown vector is composed of the coefficients u_i and q_i associated to the Neumann and Dirichlet boundary conditions, respectively. However, T-complete functions occur in the boundary integral formulation and play the same role as in the fundamental solution for the BIEM.

Boundary integral formulation.   By projecting the Laplace equation on function $\hat{u}$ that is assumed to have enough regularity, and integrating by parts twice, we get

$$0 = \int_{\Gamma} \left(q \, \hat{u} - u \, \nabla_n \hat{u} \right) \mathrm{d}x + \int_{\Omega} \left(u \, \nabla^2 \hat{u} \right) \mathrm{d}x.$$

To vanish the domain integral term, assume that $\hat{u}$ is expanded by T-complete functions associated to the Laplace operator:

$$\hat{u}(x) = \sum_{i=1}^{N} a_i \, u_i^{\star}(x) = a \cdot u^{\star}(x),$$

where the T-functions evaluated at a given point $x = \rho e^{i\theta}$ are

$$u^{\star} = \left\{ 1, \ldots, \left(\rho e^{i\theta} \right)^k, \ldots, \left(\rho e^{i\theta} \right)^n \right\}.$$

Thus, a straightforward substitution of the expression for $\hat{u}$ yields

$$0 = \int_\Gamma \left[q \left(a \cdot u^\star(x) \right) - u \left(a \cdot q^\star(x) \right) \right] dx \,.$$

This time, by substituting u and q with their approximations $\tilde{u}$ and $\tilde{q}$, we get

$$0 = \int_\Gamma \left[(\Phi(x) \cdot \mathbf{q}) \left(a \cdot u^\star(x) \right) - (\Phi(x) \cdot \mathbf{u}) \left(a \cdot q^\star(x) \right) \right] dx$$

$$= \int_\Gamma \left[\left(a \cdot u^\star \right) (\Phi \cdot \mathbf{q}) - \left(a \cdot q^\star \right) (\Phi \cdot \mathbf{u}) \right] dx.$$

Thus, by left-*factorizing* the vector a, we get N sufficient conditions for holding the above-mentioned boundary integral equation:

$$0 = \int_\Gamma \left[u^\star \, \Phi^\top \right] \mathbf{q} \, dx \; - \int_\Gamma \left[q^\star \, \Phi^\top \right] \mathbf{u} \, dx \,.$$

The integrand terms in $[\cdot]$ are matrices of dimension $N \times M$.

Matrix system resolution. The elements of matrices $\mathbf{G}$ and $\mathbf{H}$ are defined by

$$G_{ij} = \int_\Gamma \left[u_i^\star \phi_j \right] dx \qquad\qquad H_{ij} = - \int_\Gamma \left[q_i^\star \phi_j \right] dx \,.$$

In addition, the N equations obtained from the boundary integral formulation read as

$$\mathbf{G}\,\mathbf{u} + \mathbf{H}\,\mathbf{q} = 0.$$

Moreover, we can split each vector $\mathbf{u}$ and $\mathbf{q}$ involving the parts associated to the Dirichlet and Neumann boundary conditions as

$$\mathbf{u} = {}^\top (\bar{u}, \, \mathbf{u}_N) \,, \qquad\qquad \mathbf{q} = {}^\top (\mathbf{q}_D, \, \Upsilon) \,,$$

and the matrix system becomes

$$[\mathbf{G}_D, \mathbf{G}_N] \begin{pmatrix} \bar{u} \\ \mathbf{u}_N \end{pmatrix} + [\mathbf{H}_D, \mathbf{H}_N] \begin{pmatrix} \mathbf{q}_D \\ \Upsilon \end{pmatrix} = 0 \,.$$

This can be recast by moving all the unknowns to the LHS and the known boundary information to the RHS as follows:

$$[\mathbf{H}_D, \mathbf{G}_N] \begin{pmatrix} \mathbf{q}_D \\ \mathbf{u}_N \end{pmatrix} = - [\mathbf{G}_D, \mathbf{H}_N] \begin{pmatrix} \bar{u} \\ \Upsilon \end{pmatrix}.$$

7.2.4   Final Remarks

The boundary integral approaches, also called *meshless* techniques, present great conceptually intrinsic advantages: the number of degrees of freedom is drastically reduced, and thus the computational cost may theoretically be suitable for simulation over large spatial 3D domains (e.g. for transfer into the soil). However, these approaches seem to have severe limitations when applied to nonacademic cases: the implementation of Robin-type boundary conditions or the inclusion of a source term induces specific numerical treatments that can significantly degrade the gain of an approach. More fundamentally, Green functions and Trefftz-complete functions are available only for few types of equations that are linear. The consequence of this limitation is that these techniques can scarcely be applied to *realistic* configurations in which the physical properties of the medium are not constant. However, *meshless* approaches will still be used in the future; coupled with other methods (e.g. domain decomposition and multigrid methods), their use can be of great interest in building physics applications.

Chapter 8
Spectral Methods

A numerical simulation is like sex.
If it is good, then it is great.
If it is bad, then it is still better than nothing.
— Dr. D (paraphrasing Dr. Z)
People think they do not understand Mathematics,
but it is all about how you explain it to them.
— I. M. Gelfand

This chapter is organized as follows. First, we present some theoretical bases behind spectral discretizations in Sect. 8.1. An application to a problem stemming from the building physics is given in Sect. 8.3. Finally, we give some indications for the further reading in Sect. 8.4. This document contains also a certain number of Appendices directly or indirectly related to spectral methods. For instance, in Appendix 8.5 we give some useful identities about Tchebyshev polynomials and in Appendix 8.6 we give some flavour of Trefftz methods, which are essentially forgotten nowadays. Finally, we prepared also an Appendix 8.7 devoted to the Monte–Carlo methods to simulate numerically diffusion processes. The Authors admit that it is not related to spectral methods, but nevertheless we decided to include it since these methods remain essentially unknown in the community of numerical methods for Partial Differential Equations (PDEs).

8.1 Introduction to Spectral Methods

Consider for simplicity a 1D compact[1] domain, e.g. $\mathscr{U} = [-1, 1]$ an Ordinary or Partial Differential Equation (ODE or PDE) on it

---

[1] A domain $\mathscr{U} \subseteq \mathbb{R}^d$, $d \geq 1$ is compact if it is bounded and closed. For a more general definition of compactness we refer to any course in General Topology.

© Springer Nature Switzerland AG 2019
N. Mendes et al., *Numerical Methods for Diffusion Phenomena in Building Physics*,
https://doi.org/10.1007/978-3-030-31574-0_8

$$\mathcal{L}u = g, \qquad x \in \mathcal{U} \tag{8.1}$$

where $u(x, t)$ (or just $u(x)$ in the ODE case) is the solution[2] which satisfies some additional (boundary) conditions at $x = \pm 1$ depending on the (linear or nonlinear) operator $\mathcal{L} = \mathcal{L}(\partial_t, \partial_x, \partial_{xx}, \ldots)$. Function $g(x, t)$ is a source term which does not depend on the solution u. For example, if (8.1) is the classical heat equation in the free space (i.e. without heat sources), then

$$\mathcal{L} \equiv \partial_t - \nu\,\partial_{xx}, \qquad g \equiv 0,$$

where $\nu \in \mathbb{R}$ is the diffusion coefficient. If the solution $u(x)$ is steady (i.e. time-independent), we deal with an ODE. In any case, in the present document we focus only on the discretization in space. The time discretization is discussed deeper in Chap. 3. A brief reminder of some most useful numerical techniques for ODEs is given in Appendix 8.9.

The idea behind a spectral method is to approximate a solution $u(x, t)$ by a finite sum

$$u(x, t) \approx u_n(x, t) = \sum_{k=0}^{n} v_k(t)\,\phi_k(x), \tag{8.2}$$

where $\{\phi_k(x)\}_{k=0}^{\infty}$ is the set of basis functions. In ODE case all $v_k(t) \equiv$ const. The main question which arises is how to choose the basis functions? Once the choice of $\{\phi_k(x)\}_{k=0}^{\infty}$ is made, the second question appears: how to determine the expansion coefficients $v_k(t)$?

Here we shall implicitly assume that the function $u(x, t)$ is *smooth*. Only in this case the full potential of spectral methods can be exploited.[3] The concept of *smoothness* in Mathematics is ambiguous since there are various classes of smooth functions:

$$C^p(\mathcal{U}) \supseteq C^{\infty}(\mathcal{U}) \supseteq \mathcal{A}^{\infty}(\mathcal{D}), \qquad (p \in \mathbb{Z}^+, \ \mathcal{U} \subseteq \mathcal{D} \subseteq \mathbb{C}).$$

The last sequence of inclusions needs perhaps some explanations:

$C^p(\mathcal{U})$ the class of functions $f : \mathcal{U} \mapsto \mathbb{R}$ having at least $p \geq 1$ continuous derivatives.

$C^{\infty}(\mathcal{U})$ the class of functions $f : \mathcal{U} \mapsto \mathbb{R}$ having infinitely many continuous derivatives.

---

[2]For simplicity we consider in this section scalar equations only. The generalizations for systems is straightforward, since one can apply the same discretization for every individual component of the solution.

[3]We have to say that pseudo-spectral methods can be applied to problems featuring eventually discontinuous solutions as well (e.g. hyperbolic conservation laws). However, it is out of scope of the present lecture devoted rather to parabolic problems. As a side remark, the Author would add that the advantage of pseudo-spectral methods for nonlinear hyperbolic equations is not so clear comparing to modern high-resolution shock-capturing schemes [125].

$\mathscr{A}^{\infty}(\mathscr{D})$ the class of functions $f : \mathscr{D} \mapsto \mathbb{C}$ analytical (holomorphic) in a domain $\mathscr{D} \subseteq \mathbb{C}$ containing the segment $\mathscr{U}$ in its interior.

For example, the Runge[4] function $f(x) = \dfrac{1}{1 + 25\,x^2}$ is in $C^{\infty}([-1,\ 1])$, but not analytical in the complex plain $\mathscr{A}^{\infty}(\mathbb{C})$.

8.1.1 Choice of the Basis

A successful expansion basis meets the following requirements:

1. [**Convergence**] The approximations $u_n(x,\ t)$ should converge rapidly to $u(x,\ t)$ as $n \to \infty$
2. [**Differentiation**] Given coefficients $\{v_k(t)\}_{k=0}^{n}$, it should be easy to determine another set of coefficients[5] $\{v_k'(t)\}_{k=0}^{n}$ such that

$$\frac{\partial u_n}{\partial x} = \sum_{k=0}^{n} v_k(t)\,\frac{\mathrm{d}\phi_k(x)}{\mathrm{d}x} \rightsquigarrow \sum_{k=0}^{n} v_k'(t)\,\phi_k(x).$$

3. [**Transformation**] The computation of expansion coefficients $\{v_k\}_{k=0}^{n}$ from function values $\{u(x_i,\ t)\}_{i=0}^{n}$ and the reconstruction of solution values in nodes from the set of coefficients $\{v_k\}_{k=0}^{n}$ should be easy, i.e. the conversion between two data sets is algorithmically efficient

$$\{u(x_i,\ t)\}_{i=0}^{n} \leftrightarrows \{v_k\}_{k=0}^{n}.$$

8.1.1.1 Periodic Problems

For periodic problems it is straightforward to propose a basis which satisfies the requirements (1)–(3) above. It consists of *trigonometric polynomials*:

$$u_n(x,\ t) = a_0(t) + \sum_{k=1}^{n} \{a_k(t)\,\cos(k\pi x) + b_k(t)\,\sin(k\pi x)\}. \tag{8.3}$$

The first two points are explained in elementary courses of analysis, while the requirement (3) is possible thanks to the invention of the Fast Fourier Transform (FFT) algorithm first by Gauß and later by Cooley and Tukey in 1965 [42].

---

[4]Carl David Tolmé Runge (1856–1927) is a German Mathematician and Physicist. A Ph.D. student of Karl Weierstrass.

[5]Here the prime does not mean a time derivative!

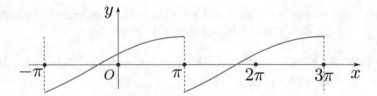

Fig. 8.1 Periodisation of a smooth continuous function defined on $[-\pi,\ \pi]$

Remark 8.1 The use of trigonometric bases such as (8.3) or $\{e^{ik\pi x}\}_{k \in \mathbb{Z}}$ for heat conduction problems has been initiated by Fourier[6] [66] even if Fourier series have been known well before Fourier. There is the so-called Arnold's[7] principle which states that in Mathematics nothing is named after its true inventor. The question the Author would like to rise is whether Arnold's principle is applicable to it-self?

Remark 8.2 The term 'Spectral methods' can be now explained. It comes from the fact that the solution $u(x,\ t)$ is expanded into a series of orthogonal eigenfunctions of some linear operator $\mathscr{L}$ (with partial or ordinary derivatives). In this way, the numerical solution is related to its spectrum, thus justifying the name 'spectral methods'. For example, if we take the Laplace operator $\mathscr{L} = -\nabla^2 \equiv -\sum_{j=1}^{d} \dfrac{\partial^2}{\partial x_j^2}$ on the periodic domain $[\,0,\ 2\pi)^d$, its spectrum consists of Fourier modes:

$$-\nabla^2 e^{-i\,k\cdot x} = |\,k\,|^2\,e^{-i\,k\cdot x}\,.$$

In this way we obtain naturally the Fourier analysis and Fourier-type pseudo-spectral methods.

8.1.1.2   Non-periodic Problems

The trigonometric basis (8.3) fails to work for general non-periodic problems essentially because of the failure of requirement (1). Indeed, the artificial discontinuities arising after the periodisation (see Fig. 8.1) make the Fourier coefficients v_n decay as $\mathscr{O}(n^{-1})$ when $n \to \infty$.

The analytic series (or Taylor-like expansions) represent another interesting alternative:

$$u_n(x,\ t) = \sum_{k=1}^{n} v_k(t)\,x^k\,. \tag{8.4}$$

---

[6]Jean-Baptiste Joseph Fourier (1768–1830) is a French Physicist and Mathematician. In particular he accompanied Napoleon Bonaparte on his campaign to Egypt as a scientific adviser.

[7]Vladimir Arnold (1937–2010), a prominent Soviet/Russian mathematician. Please, read his books!

The problem is that Taylor-type expansions (8.4) work well only for a very limited class of functions. Namely, they satisfy the requirement (1) only for functions analytical[8] in the unit disc $\mathscr{D}_1(0) \subseteq \mathbb{C}$. For instance, the celebrated Runge's function $f(x) = \dfrac{1}{1 + 25x^2}$ fails to satisfy this condition because of two imaginary poles located at $z = \pm\dfrac{i}{5}$.

The successful examples of polynomial bases are given by *orthogonal polynomials*, which arise naturally in various contexts:

Numerical integration   Optimal numerical integration formulas achieve a high accuracy by using zeros of orthogonal polynomials as nodes.

Sturm–Liouville problem   Jacobi[9] polynomials arise as eigenfunctions to singular Sturm[10]–Liouville[11] problems.

Approximation in L^2 Truncated expansions in Legendre[12] polynomials are optimal approximants in the L^2-norm.

Approximation in L^∞ Truncated expansions in Tchebyshev[13] polynomials are optimal approximants in the L^∞-norm.

Each of the topics above deserves a separate course to be covered. Here we content just to provide this information as facts, which can be deepened later, if necessary. We would just like to quote Bernardi and Maday [21]:

> We do think that the corner stone of collocation techniques is the choice of the collocation nodes […] in spectral methods these are always built from the nodes of a Gauß quadrature formula.

Consequently, extrema (and zeros) of Tchebyshev (and some other orthogonal) polynomials play a very important rôle in the Numerical Analysis (NA). Tchebyshev nodes are given explicitly by

$$x_k = -\cos\left(\frac{\pi k}{N}\right) \in [-1, 1], \qquad k = 0, 1, 2, \ldots, N. \qquad (8.5)$$

There is a result saying that using nodes (8.5) as interpolation points gives an interpolant which is not too far from the optimal polynomial $\mathscr{P}_N^{\mathrm{Opt}}$, i.e.

$$\| f - \mathscr{P}_N^{\mathrm{Tch}} \|_\infty \leq \left(1 + \Lambda_N^{\mathrm{Tch}}\right) \| f - \mathscr{P}_N^{\mathrm{Opt}} \|_\infty,$$

---

[8]The analyticity property is understood here in the sense of the Complex Analysis.

[9]Carl Gustav Jacob Jacobi (1804–1851) is a German Mathematician.

[10]Jacques Charles François Sturm (1803–1855) is a French Mathematician who was born in Geneva which was a part of France at that time.

[11]Joseph Liouville (1809–1882) is a French Mathematician who founded the *Journal de Mathmatiques Pures et Appliquées*.

[12]Adrien-Marie Legendre (1752–1833) is a French Mathematician. Not to be confused with an obscure politician Louis Legendre. Their portraits are often confused.

[13]Pafnuty Tchebyshev (1821–1894), a Russian mathematician who contributed to many fields of Mathematics from number theory to probabilities and numerical analysis.

Fig. 8.2 Interpolation of the Runge function
$$f(x) = \frac{1}{1+25x^2} \text{ on a}$$
sequence of successively refined of uniform grids. One can observe the divergence phenomenon close to the interval boundaries $x = \pm 1$

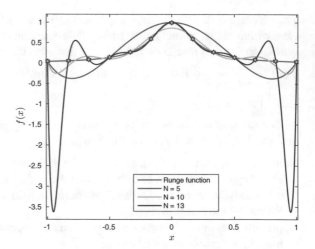

where $f : [-1, 1] \mapsto \mathbb{R}$ is the function that we interpolate and $\mathscr{P}_N^{\text{Tch}}$ is the interpolation polynomial constructed on nodes (8.5). Here Λ_N^{Tch} is the so-called Lebesgue[14] constant for Tchebyshev interpolation polynomials of degree N. Notice that Λ_N^{Tch} does not depend on the function f being interpolated. The Lebesgue constants for various interpolation techniques have the following asymptotic behaviour

$$\Lambda_N^{\text{Tch}} \sim \mathcal{O}(\log N), \qquad \Lambda_N^{\text{Leg}} \sim \mathcal{O}(\sqrt{N}), \qquad \Lambda_N^{\text{Uni}} \sim \mathcal{O}\left(\frac{2^N}{N \log N}\right),$$

where Λ_N^{Leg} and Λ_N^{Uni} are Lebesgue constants for Legendre and uniform nodes distributions correspondingly. In particular one can see that the uniform node distribution is simply disastrous. This phenomenon is another 'avatar' of the so-called Runge phenomenon illustrated in Fig. 8.2. The performance of Tchebyshev's nodes is shown in Fig. 8.3.

Remark 8.3 As it was shown by Vértesi [194], the Lebesgue constant for Tchebyshev distribution of nodes is very close for the smallest possible Lebesgue constant:

$$\Lambda_N^{\text{Tch}} = \frac{2}{\pi} \left(\ln N + \gamma + \ln \frac{8}{\pi} \right) + o(1),$$

$$\Lambda_N^{\text{min}} = \frac{2}{\pi} \left(\ln N + \gamma + \ln \frac{4}{\pi} \right) + o(1).$$

---

[14]Henri Lebesgue (1875–1941), a French mathematician most known for the theory of integration having its name.

Fig. 8.3 Interpolation of the Runge function
$$f(x) = \frac{1}{1 + 25x^2} \text{ on a}$$
sequence of successively refined of Tchebyshev's grids. One can see that the oscillations present in Fig. 8.2 disappear and the interpolant seems to converge to the interpolated function $f(x)$

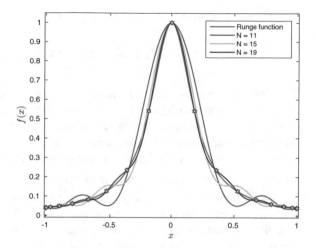

where $\gamma \approx 0.5772156649\ldots$ is the Euler[15]–Mascheroni[16] constant.

The Lebesgue Constant

It is worth to explain better the important notion of the Lebesgue constant and how it appears in the theory of interpolation (below we follow [183]). Let us take an arbitrary (valid) nodes distribution $\{x_i\}_{i=0}^{n}$ in domain $\mathscr{U}$, i.e.

$$\forall i = 1, 2, \ldots, n : x_i \in \mathscr{U}, \qquad \forall j \neq i : x_i \neq x_j.$$

For any continuous function $u \in C(\mathscr{U})$ there exists a unique interpolating polynomial $\mathscr{P}(x) \in \mathbb{P}_n[\mathbb{R}]$ of degree $n = \deg \mathscr{P}$:

$$\mathscr{P}_n(x_i) = u_i \equiv u(x_i), \qquad i = 1, 2, \ldots, n.$$

This interpolating polynomial $\mathscr{P}_n(x)$ will be denoted as $\mathbb{I}_n[u]$. We introduce also the best possible approximating polynomial $\mathscr{P}^* \in \mathbb{P}_n[\mathbb{R}]$:

$$\| u - \mathscr{P}^* \|_\infty = \inf_{\mathscr{P} \in \mathbb{P}_n[\mathbb{R}]} \| u - \mathscr{P} \|_\infty.$$

It is not obliged that $\mathscr{P}^* \equiv \mathbb{I}_n[u]$. However, since $\mathscr{P}^* \in \mathbb{P}_n[\mathbb{R}]$ we necessarily have $\mathscr{P}^* \equiv \mathbb{I}_n[u]$. Therefore, the following inequalities hold:

$$\| u - \mathbb{I}_n[u] \|_\infty = \| u - \mathscr{P}^* + \mathscr{P}^* - \mathbb{I}_n[u] \|_\infty \equiv$$
$$\| u - \mathscr{P}^* + \mathbb{I}_n[\mathscr{P}^*] - \mathbb{I}_n[u] \|_\infty \leq \| u - \mathscr{P}^* \|_\infty + \| \mathbb{I}_n[\mathscr{P}^*] - \mathbb{I}_n[u] \|_\infty$$
$$\leq \| u - \mathscr{P}^* \|_\infty + \| \mathbb{I}_n \| \cdot \| \mathscr{P}^* - u \|_\infty \leq \left(1 + \| \mathbb{I}_n \| \right) \cdot \| u - \mathscr{P}^* \|_\infty.$$

---

[15]Leonhard Euler (1707–1783) is a great mathematician who was born in Switzerland and worked all his life in Saint-Petersburg.

[16]Lorenzo Mascheroni (1750–1800) was an Italian mathematician who worked in Pavia.

The norm of a linear operator $\mathbb{I}_n [\cdot]$ is defined as

$$\| \mathbb{I}_n \| \overset{\text{def}}{:=} \sup_{\| u \|_\infty = 1} \| \mathbb{I}_n [u] \|_\infty .$$

The aforementioned norm of the interpolation operator $\mathbb{I}_n [\cdot]$ is called the Lebesgue constant for the set of nodes $\{x_i\}_{i=0}^n$. In multiple dimensions the Lebesgue constant depends also on the shape of domain $\mathscr{U}$ additionally to the nodes distribution $\{x_i\}_{i=0}^n$. As we mentioned above, obvious choices such as the uniform distribution of nodes is disastrous, since it yields the exponential growth of the Lebesgue constant $\Lambda_n^{\text{Uni}} \sim \mathscr{O}\left(\frac{2^n}{n \log n}\right)$, where n is the degree of the interpolating polynomial.

The estimation of the Lebesgue constant in multi-dimensional non-Cartesian domains is a problem essentially open nowadays. The most precious information is to find the nodes distribution, for example over a triangle, which minimizes the Lebesgue constant. This knowledge would be crucial for the design of new spectral elements [183]. The locations of nodes which minimize the magnitude of the Lebesgue constant $\Lambda_n \to$ min for a given polynomial space $\mathbb{P}_n [\mathbb{R}]$ are called Lebesgue points. Nowadays, the best known points on quadrangles are tensor products of Gauß–Lobatto[17] or Tchebyshev nodes. On triangles the Fekete[18] points seem currently to be the best choice.

Intermediate Conclusions

The remarks above show that Tchebyshev polynomials satisfy the requirement (1) for a useful spectral basis. The requirement (2) is met due to derivative recursion formulas, which can be easily demonstrated (see also Appendix 8.5):

$$T_k'(x) = 2k \sum_{n=0}^{\left[\frac{k-1}{2}\right]} \frac{1}{\delta_{k-1-2n}} T_{k-1-2n}(x), \qquad \delta_k = \begin{cases} 2, & k = 0, \\ 1, & k \neq 0. \end{cases}$$

Finally, the requirement (3) is satisfied as well thanks to the Fast Cosine Fourier Transform (FCFT) (a variant of FFT). It allows to compute spectral coefficients $\{v_k\}_{k=0}^N$ from the node values $\{u_n(x_i)\}_{i=0}^N$ and vice versa. Consequently, Tchebyshev polynomials have become an almost universal choice for non-periodic problems. These methods in 1D have been implemented in the Matlab toolbox Chebfun[19] (and Chebfun2 in 2D).

Remark 8.4 To Authors' knowledge, Legendre polynomials found some applications in the construction of Spectral Element Method (SEM) bases [183].

---

[17]Rehuel Lobatto is a Dutch Mathematician born in a Portuguese family.

[18]Michael Fekete (1886–1957) is a Hungarian Mathematician who did his Ph.D. under the supervision of Lipót Fejér. M. Fekete gave also some private tutoring to JánosNeumann known today as John von Neumann.

[19]Chebfun's team is lead by Nick Trefethen.

Remark 8.5 We have to mention that high-order polynomial interpolants have a bad reputation in the Numerical Analysis (NA) community for the two following reasons:

1. Because of the Runge phenomenon
2. and because of the following

Theorem 8.1 *For any node density distribution function there exists a continuous function such that the L^∞-norm of the interpolation error tends to infinity when the number of nodes $N \to +\infty$.*

However, this bad reputation is not justified. For instance, the Runge phenomenon is completely suppressed by Tchebyshev nodes distribution. The low or moderate Lebesgue constant Λ_N ensures that we are not too far from the optimal polynomial.

Infinite Domains

The infinite domains can be handled, first of all, by periodisation in the context of Fourier-type methods as explained above. However, truly infinite domains can be handled by various approaches:

- Hermite polynomials
- Sinc functions
- (almost) Rational functions
- Change of variables, e.g.

$$x : [-\pi, \pi] \mapsto \mathbb{R}, \qquad x(q) = \ell \tan\left(\frac{q}{2}\right), \qquad \ell \in \mathbb{R}^+$$

- Truncation of a domain $(-\infty, \infty) \rightsquigarrow [-\ell, \ell]$ followed by a change of variables such as

$$x : [-\pi, \pi] \mapsto [-\ell, \ell], \qquad x(q) = \frac{\ell q}{\pi}.$$

Two last items are followed by a conventional pseudo-spectral discretization on a finite domain $[-\pi, \pi]$.

A Hermite[20]-type pseudo-spectral method can be briefly sketched as follows. First of all, we construct recursively the family of Hermite's polynomials:

$$\mathcal{H}_0(x) = 1, \qquad \mathcal{H}_1(x) = 2x, \qquad \mathcal{H}_{n+1}(x) = 2x\,\mathcal{H}_n(x) - 2n\,\mathcal{H}_{n-1}(x), \qquad n \geq 1.$$

Then, we construct Hermite's functions, which form an *orthonormal* basis on $\mathbb{R}$:

$$\mathcal{H}_n(x) \stackrel{\text{def}}{:=} \frac{1}{\sqrt{2^n \cdot n!}}\, \mathcal{H}_n(x)\, e^{-\frac{x^2}{2}}.$$

---

[20]Charles Hermite (1822–1901), a French mathematician whose *"Cours d'analyse"* represent a lot of interest even today. He was also the Ph.D. advisor of another great French mathematician—Henri Poincaré.

Derivatives of expansions in Hermite functions can be easily re-expanded using the relation:

$$\mathscr{H}_n'(x) = -\sqrt{\frac{n+1}{2}}\,\mathscr{H}_{n+1}(x) + \sqrt{\frac{n}{2}}\,\mathscr{H}_{n-1}(x).$$

However, the lack of a Fast Hermite Transform (FHT) implies the use of *differentiation matrices* in numerical implementations. There are also some concerns about a poor convergence rate when the number of modes N is increased. The infinite domains remain very challenging *in silico*.

8.1.2 Determining Expansion Coefficients

There are three main techniques to find the spectral expansion coefficients $\{v_k\}_{k=0}^N$. In order to explain them, let us introduce first the residual function on a trial solution $u_n(x, t)$:

$$\mathscr{R}[u_n](x, t) \stackrel{\text{def}}{:=} \left[\mathscr{L}u_n - g\right](x, t).$$

Typically, we shall evaluate the residual $\mathscr{R}$ on the expansion (8.2) and the residual norm $\|\mathscr{R}\|$ is generally considered as a measure of the approximate solution quality. The goal is to keep the residual as small as possible across the domain $\mathscr{U}$ (in our particular case $\mathscr{U} = [-1, 1]$).

So, we can mention here at least five approaches to determine the expansion coefficients:

Tau–Lanczos   Spectral coefficients $\{v_k\}_{k=0}^N$ are selected such that the boundary conditions are satisfied identically and the residual $\mathscr{R}[u_n]$ is orthogonal to as many basis functions $\phi_k(x)$ as possible.

Galerkin   First the basis functions are recombined $\{\phi_k(x)\}_{k=0}^N \rightsquigarrow \{\tilde{\varphi}_k(x)\}_{k=0}^N$ so that the boundary conditions are satisfied identically. Then, the coefficients $\{v_k\}_{k=0}^N$ are found so that the residual $\mathscr{R}[u_n]$ be orthogonal to as many of *new* basis functions $\{\tilde{\phi}_k(x)\}_{k=0}^N$ as possible.

Collocation   This approach is similar to the Tau–Lanczos method concerning the boundary conditions: spectral coefficients $\{v_k\}_{k=0}^N$ are selected such that the boundary conditions are satisfied. The rest of coefficients is determined so that the residual $\mathscr{R}[u_n](x, t)$ vanishes at as many (thoroughly chosen) spatial locations as possible.

Petrov–Galerkin   It is a variant of Galerkin method in which the residual $\mathscr{R}[u_n]$ is made orthogonal to a set of functions, which is different from the approximation space basis $\{\phi_k(x)\}_{k=0}^N$.

Least squares   Various least square-type approaches are used when for some reason the number of coefficients to be determined is different from the number of conditions which can be imposed. Below we shall avoid such pathological situations.

The Tau–Lanczos technique was proposed by C. Lanczos[21] in 1938. What we call the Galerkin[22] technique was proposed independently first by I. Bubnov[23] and by W. Ritz[24] (one more example of the Arnold principle in action!). So, to respect the historical time line, the method should be called Bubnov–Ritz–Galerkin. Today it is the basis of the Finite Element Method (FEM). The collocation technique was called the *pseudo-spectral method* presumably for the first time by S. Orszag[25] in 1972. The Petrov–Galerkin method was proposed by G.I. Petrov.[26] It is used up to now in some convection-dominated problems.

Remark 8.6 The collocation method can be recast into the Petrov–Galerkin framework when we make the residual $\mathscr{R}[u_n]$ is made orthogonal to Dirac[27] singular measures $\left\{\delta(x - x_k)\right\}_{k=1}^{n}$, where $\{x_k\}_{k=1}^{n}$ are the collocation points.

For linear problems all these methods work equally well. However, for nonlinear ones (and in the presence of variable coefficients) the pseudo-spectral (collocation) approach is particularly easy to apply since it involves the products of numbers (solution/variable coefficient's values in collocation points) instead of products of expansions, which are much more difficult to handle.

The convergence of pseudo-spectral approximations for very smooth functions is always geometrical, i.e. $\sim\mathscr{O}(q^N)$, where N is the number of modes. This statement is true for any derivative with the same convergence factor $0 < q < 1$. However, the periodic pseudo-spectral method converges always faster than its non-periodic counterpart. This conclusion follows from convergence properties of Fourier and Tchebyshev series in the complex domain.

The relative resolution ability of various pseudo-spectral methods can be also quantified in terms of the number of points per wavelength needed to resolve a signal. Indeed, this description is more suitable for wave propagation problems. For periodic Fourier-type methods one needs 2 points per wavelength. For Tchebyshev-type methods one needs about π points. Finally, this number goes up to 6 nodes per wavelength for uniform grids. However, due to the huge Lebesgue constant Λ_N^{Uni} the uniform grids in pseudo-spectral setting are not usable in practice.

---

[21]Cornelius Lanczos (1893–1974), a Hungarian/American numerical analyst. His books are very pedagogical as well.

[22]Boris Galerkin (more precisely romanized as Galyorkin) (1871–1945), a Russian then Soviet Civil Engineer. The Author suggests to read his biography which is comparable to James Bond (007) movies.

[23]Ivan Bubnov (1872–1919), a Russian naval engineer.

[24]Walther Ritz (1878–1909), a talented Swiss Physicist, who died young in Göttingen, Germany.

[25]Steven Orszag (1943–2011), an American numerical analyst, one of the first users of pseudo-spectral methods.

[26]Georgii Ivanovich Petrov (1912–1987), a Russian fluid mechanician.

[27]Paul Adrien Maurice Dirac (1902–1984) is a British Theoretical Physicist who predicted theoretically using Dirac's equation the existence of the positron. One of the founders of the Quantum Mechanics.

8.2 Aliasing, Interpolation and Truncation

Let us take a continuous (and possibly a smooth) function $u(x)$ defined on the interval $\mathscr{I} = (-\pi, \pi)$ and develop it in a Fourier series. In general it will contain the whole spectrum (i.e. the infinite number) of frequencies:

$$u(x) = \sum_{k=-\infty}^{+\infty} v_k\, e^{ikx}.$$

Now let us discretize the interval $\mathscr{I}$ with N equispaced collocation points (as we do it in Fourier-type pseudo-spectral methods. In the following we assume N to be odd, i.e. $N = 2m + 1$. On this discrete grid all modes $\left\{ e^{i(k+jN)x} \right\}_{j\in\mathbb{Z}}$ are indistinguishable. See Fig. 8.4 for an illustration of this phenomenon.

The interpolating trigonometric polynomial on a given grid can be written as

$$\mathbb{I}_N[u] = \sum_{k=-m}^{m} \hat{v}_k\, e^{ikx}.$$

Each discrete Fourier coefficient incorporates the contributions of all modes which looks the same on the considered grid:

$$\hat{v}_k = \sum_{j=-\infty}^{+\infty} v_{k+jN}.$$

Let us recall that the polynomial $\mathbb{I}_N[u]$ takes the prescribed values $\left\{ u(x_k) \right\}_{k=-m}^{m}$ in the points of the grid $\{x_k\}_{k=-m}^{m}$. This object is fundamentally different from the truncated Fourier series:

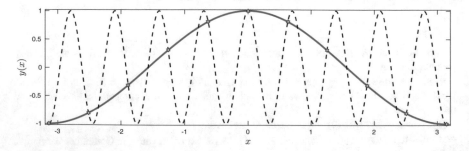

Fig. 8.4 Illustration of the aliasing phenomenon: two Fourier modes are indistinguishable on the discrete grid. The modes represented here are $\cos(x)$ and $\cos(9x)$ and the discrete grid is composed of $N = 11$ equispaced points on the segment $[-\pi, \pi]$

$$\mathbb{T}_N[u] = \sum_{k=-m}^{m} v_k \, e^{ikx} .$$

The difference between these two quantities is known as the *aliasing error*:

$$\mathscr{R}_N[u] \overset{\text{def}}{:=} \mathbb{I}_N[u] - \mathbb{T}_N[u] = \sum_{k=-m}^{m} \sum_{\forall j \neq 0} v_{k+jN} \, e^{ikx} .$$

After applying the Pythagoras theorem,[28] we obtain

$$\| u - \mathbb{I}_N[u] \|_{L_2}^2 = \| u - \mathbb{T}_N[u] \|_{L_2}^2 + \| \mathscr{R}_N[u] \|_{L_2}^2 .$$

Thus, the interpolation error is always larger than the truncation error in the standard L^2 norm. The amount of this difference is precisely equal to the committed *aliasing error*. However, we prefer to use in pseudo-spectral methods the interpolation technique because of the Discrete Fourier Transform (DFT), which allows to transform quickly (thanks to the FFT algorithm) from the set of function values in grid points to the set of its interpolation coefficients. So, it is easier to apply an FFT instead of computing N integrals to determine the Fourier series coefficients.

Nonlinearities

Let us take the simplest possible nonlinearity—the product of two functions $u(x)$ and $v(x)$ defined by their truncated Fourier series containing the modes up to m :

$$u(x) = \sum_{k=-m}^{m} u_k \, e^{ikx} , \qquad v(x) = \sum_{k=-m}^{m} v_k \, e^{ikx} .$$

The product of these two functions $w(x)$ can be obtained my multiplying the Fourier series:

$$w(x) = u(x) \cdot v(x) = \left(\sum_{k=-m}^{m} u_k \, e^{ikx} \right) \cdot \left(\sum_{k=-m}^{m} v_k \, e^{ikx} \right) \equiv \sum_{k=-2m}^{2m} w_k \, e^{ikx} .$$

It can be clearly seen that the product contains high order harmonics up to $e^{\pm imx}$ which cannot be represented on the initial grid. Thus, they will contribute to the aliasing error explained above.

The aliasing of a nonlinear product can be ingeniously avoided by adopting the so-called 3/2th rule whose Matlab implementation is given below. This function assumes that input vectors are Fourier coefficients of functions $u(x)$ and $v(x)$. The

---

[28]We can apply the Pythagoras theorem since the aliasing error $\mathscr{R}_N[u]$ contains the Fourier modes with numbers $|k| \leq m$, while the reminder $u - \mathbb{T}_N[u]$ contains only the modes with $|k| > m$. Thus, they are orthogonal.

resulting vector contains (anti-aliased) Fourier coefficients of their product $w(x) = u(x) \cdot v(x)$.

```
 1  function w_hat = AntiAlias(u_hat, v_hat)
 2    N          = length(u_hat);
 3    M          = 3*N/2; % 3/2th rule
 4    u_hat_pad  = [u_hat(1:N/2) zeros(1, M-N) u_hat(N/2+1:end)];
 5    v_hat_pad  = [v_hat(1:N/2) zeros(1, M-N) v_hat(N/2+1:end)];
 6    u_pad      = ifft(u_hat_pad);
 7    v_pad      = ifft(v_hat_pad);
 8    w_pad      = u_pad.*v_pad;
 9    w_pad_hat  = fft(w_pad);
10    w_hat      = 3/2*[w_pad_hat(1:N/2) w_pad_hat(M-N/2+1:M)];
11  end % AntiAlias()
```

The main idea behind is to complete vectors of Fourier coefficients by a sufficient number of zeros (i.e. the so-called zero padding technique) so that in the physical space the product $u(x) \cdot v(x)$ can be fully resolved. The final step consists in extracting m relevant Fourier coefficients [189, 191].

Remark 8.7 To Authors's knowledge, the development of efficient and rigorously justified anti-aliasing rules for other types of nonlinearities such as the division, square root, etc. is an open problem.

8.2.1 Example of a Second Order Boundary Value Problem

Consider the following second order Boundary Value Problem (BVP) on the interval $\mathscr{I} = [-1, 1]$:

$$u_{xx} + u_x - 2u + 2 = 0, \quad u(-1) = u(1) = 0. \tag{8.6}$$

This BVP has the exact solution

$$u(x) = 1 - \frac{\sinh(2)}{\sinh(3)}\, e^x - \frac{\sinh(1)}{\sinh(3)}\, e^{-2x}, \quad x \in [-1, 1]. \tag{8.7}$$

We shall seek for the numerical solution to (8.6) in the form of a truncated Tchebyshev expansion:

$$u(x) \approx \sum_{k=0}^{4} a_k\, T_k(x), \quad \forall k : a_k \in \mathbb{R}.$$

Spectral coefficients $\{a_k\}_{k=0}^{4}$ have to be determined. The last expansion is substituted into the governing equation (8.6). There is no reason that (8.6) will be satisfied identically in every point of $\mathscr{I}$. Hence, we can measure the residual

$$\mathscr{R}(x) = (u_{xx} + u_x - 2u + 2)(x) \rightsquigarrow 0.$$

The enforcing of boundary conditions $u(-1) = u(1) = 0$ leads to two additional relations on spectral coefficients:

$$a_0 + a_1 + a_2 + a_3 + a_4 = 0,$$
$$a_0 - a_1 + a_2 - a_3 + a_4 = 0.$$

So, we have two relations coming from boundary conditions and we have five degrees of freedom $\{a_k\}_{k=0}^4$. It means that we have to impose three additional conditions to determine uniquely all spectral coefficients. Different approaches prescribe different numerical recipes.

8.2.1.1 Tau–Lanczos

We require that the residual $\mathscr{R}(x)$ be orthogonal to the first three basis functions $T_{0,1,2}(x)$. It gives us three additional relations:

$$\langle \mathscr{R}, T_k \rangle \equiv \int_{-1}^{1} \frac{\mathscr{R}(x) T_k(x)}{\sqrt{1 - x^2}} \, dx = 0, \quad k = 0, 1, 2.$$

8.2.1.2 Galerkin

We recombine the Tchebyshev polynomials to form a different basis which satisfies identically the boundary conditions:

$$\phi_0(x) \stackrel{\text{def}}{:=} (T_2 - T_0)(x),$$
$$\phi_1(x) \stackrel{\text{def}}{:=} (T_3 - T_1)(x),$$
$$\phi_2(x) \stackrel{\text{def}}{:=} (T_4 - T_0)(x).$$

Then we require that the residual $\mathscr{R}(x)$ is orthogonal to the new basis functions:

$$\langle \mathscr{R}, \phi_k \rangle \equiv \int_{-1}^{1} \frac{\mathscr{R}(x) \phi_k(x)}{\sqrt{1 - x^2}} \, dx = 0, \quad k = 0, 1, 2.$$

8.2.1.3 Collocation

This approach is particularly simple. We require that $\mathscr{R}(x_k) \equiv 0$ in three interior Tchebyshev points $x_k = \cos\left(\frac{\pi k}{4}\right), k = 1, 2, 3$.

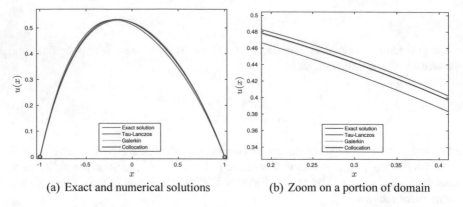

(a) Exact and numerical solutions   (b) Zoom on a portion of domain

Fig. 8.5 Comparison of various numerical approaches to determine the spectral expansion coefficients for a linear BVP (8.6)

Conclusions

The comparison of three numerical solutions with the exact one (8.7) is shown in Fig. 8.5. In this Fig. 8.5 one can see that the collocation[29] and Galerkin methods provide nearly identical[30] numerical solutions which are closer to the exact solution than the prediction of the Tau method. However, in general the Authors are impressed by the performance of spectral methods—with only five degrees of freedom $\{a_k\}_{k=0}^4$ we capture very well the global behaviour of the solution (8.7) in every point of the interval $\mathscr{I}$. Perhaps, the last point has to be emphasized again: the (pseudo-)spectral methods provide the numerical value of the solution in the *whole domain*, not only in collocation or grid points.

8.3   Application to Heat Conduction

In this section we show how to apply the Fourier-type spectral methods to simulate some simple and not so simple diffusion processes.

8.3.1   An Elementary Example

As the simplest example, consider the linear heat equation

$$u_t = \nu \, u_{xx}, \qquad x \in \mathbb{R}, \tag{8.8}$$

---

[29]The advantage of the collocation approach for nonlinear (and variable coefficients) problems becomes even more flagrant.

[30]With a little advantage towards the collocation method. However, this conclusion is not general at all. It is based only on this particular problem.

completed with the initial condition

$$u(x, 0) = u_0(x).$$

We consider this equation on the whole line $\mathbb{R}$ for the sake of simplicity. However, *in silico* it will be truncated to a periodic interval (in the example below it will be $\mathscr{I} = [-1, 1]$). Let us apply the Fourier integral transform to the both sides of Eq. 8.8:

$$\frac{\mathrm{d}\hat{u}(k, t)}{\mathrm{d}t} = -\nu k^2 \hat{u}(k, t), \tag{8.9}$$

where the forward $\hat{u}(k, t) = \mathscr{F}\{u(x, t)\}$ and inverse $u(x, t) = \mathscr{F}^{-1}\{\hat{u}(k, t)\}$ integral Fourier transforms are defined below in (8.14) and (8.15) correspondingly. The transformed heat equation (8.9) can be regarded as a linear ODE. Its solution can be readily obtained:

$$\hat{u}(k, t) = \hat{u}_0(k)\,\mathrm{e}^{-\nu k^2 t}, \qquad \hat{u}_0(k) = \mathscr{F}\{u_0(x)\}.$$

Consequently, we possess an analytical solution to the heat equation (8.8) in Fourier space. In order to obtain the solution in physical space, an inverse Fourier transform has to be computed. The exponential decay of Fourier coefficients $\hat{u}(k, t)$ (except the zero mode $k = 0$) ensures that the solution $u(x, t)$ becomes infinitely smooth C^∞ for $t > 0$. For example, the initial condition and corresponding solution at $t = T = 5\,\mathrm{s}$ are represented in Fig. 8.6. The Matlab code used to generate Fig. 8.6 is provided below as well.

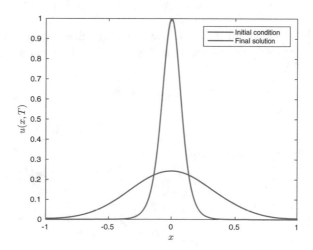

Fig. 8.6 Spectral solution to the linear heat equation (8.8) at $t = 5.0$, with $\nu = 10^{-2}$ and the initial condition is $u_0(x) = \mathrm{sech}^2(10 x)$

```
 1  l   = 1.0;        % half-length of the domain
 2  N   = 256;        % number of Fourier modes
 3  dx  = 2*l/N;      % distance between two collocation points
 4  x   = (1-N/2:N/2)*dx; % physical space discretization
 5  nu  = 0.01;       % diffusion parameter
 6  T   = 5.0;        % time where we compute the solution
 7
 8  dk = pi/l;                        % discretization step in ...
        Fourier space
 9  k   = [0:N/2 1-N/2:-1]*dk;  % vector of wavenumbers
10  k2  = k.^2;                     % almost 2nd derivative in ...
        Fourier space
11
12  u0      = sech(10.0*x).^2;  % initial condition
13  u0_hat  = fft(u0);          % Its Fourier transform
14
15  % and the solution at final time:
16  uT      = real(ifft(exp(-nu*k2*T).*u0_hat));
```

8.3.2 A Less Elementary Example

The numerical code above is based on the knowledge of the analytical solution to ODE (8.9) in the Fourier space. It is unnecessary to say that an analytical solution is available in very simple situations only. Consequently, we assume that the resulting ODE system in the physical (Fourier) space is formally solved (i.e. advanced in time) by the semigroup operator $\mathscr{S}(t)$ ($\hat{\mathscr{S}}(t) \equiv \mathscr{F} \cdot \mathscr{S}(t) \cdot \mathscr{F}^{-1}$ correspondingly):

$$u(x,\, t) \equiv \mathscr{S}(t) \cdot \big[u_0(x) \big].$$

In practice this semigroup operator is realized using numerical time-marching techniques described briefly in Appendix 8.9. So, a more general Fourier spectral algorithm is

1. Decompose the initial condition:

$$\hat{u}_0(k) = \mathscr{F}\big\{ u_0(x) \big\},$$

2. Advance in time:

$$\hat{u}_k(t) = \hat{\mathscr{S}}(t)\big[\hat{u}_0(k) \big],$$

3. Synthesize:

$$u(x,\, t) = \mathscr{F}^{-1}\big\{ \hat{u}_k(t) \big\}.$$

This algorithm is based on the fact that the following diagram commutes:

$$
\begin{array}{ccc}
u(x,\,0) & \xrightarrow{\;\mathscr{S}(t)\;} & u(x,\,t) \\[2pt]
\mathscr{F}\downarrow & & \uparrow\mathscr{F}^{-1} \\[2pt]
\hat{u}_0(k) & \xrightarrow{\;\hat{\mathscr{S}}(t)\;} & \hat{u}_k(t)
\end{array}
$$

8.3.3   A Real-Life Example

In this section we shall consider a realistic model (to be honest we take a slightly simplified version), which was proposed in [139] to predict heat and moisture transfer through the walls. This model is used in real-world Civil Engineering applications such as the code Domus. So, the model we consider in this section reads (for simplicity we consider the 1D case):

$$
\frac{\partial \theta}{\partial t} = \frac{\partial}{\partial x}\left(\mathscr{D}_\theta \frac{\partial \theta}{\partial x} + \mathscr{D}_T \frac{\partial T}{\partial x}\right), \tag{8.10}
$$

$$
\rho c_m \frac{\partial T}{\partial t} = -\frac{\partial q}{\partial x} - L(T)\cdot\frac{\partial j_v}{\partial x}, \tag{8.11}
$$

where θ is moisture volumetric content (i.e. moisture density) and T is the temperature. The mass transport coefficients $\mathscr{D}_\theta(\theta,\,T)$ and $\mathscr{D}_T(\theta,\,T)$ may depend nonlinearly on the solution $(\theta(x,\,t),\,T(x,\,t))$. Here we assume for simplicity that the mass density ρ and the specific moisture heat c_m are some positive constants.[31] Finally, the heat flux q and vapor flow j_v can be expressed as

$$
q = -\lambda(\theta,\,T)\,\frac{\partial T}{\partial x}\,,
$$

$$
j_v = -\mathscr{V}_\theta(\theta,\,T)\,\frac{\partial \theta}{\partial x} - \mathscr{V}_T(\theta,\,T)\,\frac{\partial T}{\partial x}\,.
$$

In Fourier-type pseudo-spectral methods we usually work with Fourier coefficients. Consequently, we apply the Fourier transform to both sides of Eqs. (8.10), (8.11):

$$
\frac{d\hat{\theta}}{dt} = ik\,\mathscr{F}\left\{\mathscr{D}_\theta \frac{\partial \theta}{\partial x} + \mathscr{D}_T \frac{\partial T}{\partial x}\right\}, \tag{8.12}
$$

$$
\rho c_m \frac{d\hat{T}}{dt} = ik\,\mathscr{F}\left\{\lambda(\theta,\,T)\frac{\partial T}{\partial x}\right\} - \mathscr{F}\left\{L(T)\cdot\frac{\partial j_v}{\partial x}\right\}, \tag{8.13}
$$

---

[31] In more realistic modelling $c_m(\theta)$ depends also on the moisture density θ. It does not pose any problems to take it into account in the numerical scheme described in Sect. 8.3.

where we introduced some notations for the Fourier transform $\mathscr{F}(\cdot)$:

$$\hat{\theta}(k,\,t) \overset{\text{def}}{:=} \mathscr{F}\big\{\theta(x,\,t)\big\} \;=\; \int_{-\infty}^{+\infty} \theta(x,\,t)\,\mathrm{e}^{ikx}\,\mathrm{d}x\,, \qquad (8.14)$$

where $k \in \mathbb{R}$ is the wave number. Inversely we have

$$\theta(x,\,t) \;=\; \mathscr{F}^{-1}\big\{\hat{\theta}(k,\,t)\big\} \;=\; \frac{1}{2\pi}\int_{-\infty}^{+\infty} \hat{\theta}(k,\,t)\,\mathrm{e}^{-ikx}\,\mathrm{d}k\,. \qquad (8.15)$$

Now, System (8.12), (8.13) can be considered as a coupled nonlinear system of Ordinary Differential Equations (ODEs) (not PDEs!) for the Fourier coefficients of solutions $\big(\theta(x,\,t),\,T(x,\,t)\big)$. The numerical methods for systems of ODEs were discussed in Sect. 3.1.3. As general fundamental references on this topic we can recommend [92, 93]. The Authors even suggest to employ a well-documented ready-to-use ODE libraries such as [181], for example. Now we have to explain how to evaluate the right hand side of Eqs. (8.12), (8.13) when only the Fourier coefficients are available. The recipe is very simple:

- All linear operations (e.g. additions, subtractions, multiplication by a scalar) can be equally made in Fourier or in a real spaces. The choice has to be done in order to minimize the number of FFT operations

$$\mathscr{F}\big\{\alpha\theta \,+\, \beta T\big\} \;\equiv\; \alpha\hat{\theta} \,+\, \beta\hat{T}$$

- All nonlinear products are made in the real space and then we transfer the result back to the Fourier space using one FFT, e.g.

$$\mathscr{F}\big\{\theta(x,\,t)\cdot T(x,\,t)\big\} \;=\; \mathscr{F}\big\{\mathscr{F}^{-1}\{\hat{\theta}(k,\,t)\}\cdot\mathscr{F}^{-1}\{\hat{T}(k,\,t)\}\big\}$$

- All spatial derivatives are computed only in Fourier space, as

$$\frac{\partial^n}{\partial x^n}[\,\cdot\,] \;\leftrightarrows\; \mathscr{F}^{-1}\big\{(ik)^n\,[\,\cdot\,]\big\}$$

The Reader can notice that above we used Fourier integrals. It comes from the mathematical tradition. In computer implementations one has to take a finite interval, periodize it and use direct and inverse Discrete Fourier Transforms (DFTs) instead of $\mathscr{F}\{\cdot\}$ and $\mathscr{F}^{-1}\{\cdot\}$ respectively.

The computation of derivatives using the Fourier collocation spectral method is illustrated below on the following periodic function (with period 2):

$$u(x) \;=\; \sin\big(\pi(x\,+\,1)\big)\,\mathrm{e}^{\sin(\pi(x\,+\,1))}\,, \qquad x \in [-1,\,1]\,. \qquad (8.16)$$

The first three derivatives of this function can be readily computed using any symbolic computation software (the Authors used Maple to be more specific):

$$u'(x) = \pi \cos(\pi(x + 1))\left(1 + \sin(\pi(x + 1))\right) e^{\sin(\pi(x + 1))},$$

$$u''(x) = \pi^2 \left\{\cos^2(\pi(x + 1))\left(\sin(\pi(x + 1)) + 3\right) - \sin(\pi(x + 1)) - 1\right\} e^{\sin(\pi(x + 1))},$$

$$u'''(x) = \pi^3 \cos(\pi(x + 1))\left\{\cos^2(\pi(x + 1))\left(\sin(\pi(x + 1)) + 6\right) - 7\sin(\pi(x + 1)) - 4\right\}.$$

The expressions above are used to assess the accuracy of the computed approximations in collocation points. The analytical derivatives with computed ones (red dots) are shown in Fig. 8.7 for $N = 128$ collocation points. To the graphical accuracy the results are indistinguishable. That is why we computed also the discrete ℓ_∞ norms to compute the relative errors:

$$\varepsilon_N^{(p)} = \frac{\|u_{num} - u_{exact}\|_\infty}{\|u_{exact}(x_i)\|_\infty}.$$

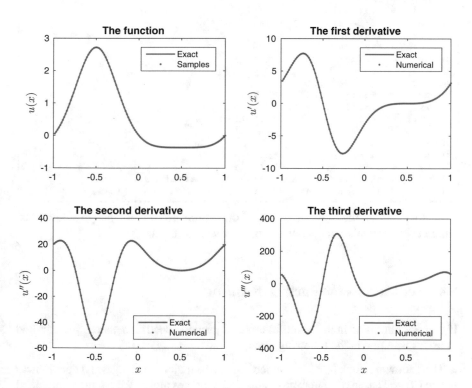

Fig. 8.7 Numerical differentiation of a periodic function (8.16) using the Fourier collocation spectral method. We use 128 collocation points. The agreement is perfect up to graphic accuracy

For $N = 32$ Fourier modes we obtain the following numerical results:

$$\varepsilon_{32}^{(1)} \approx 1.5 \times 10^{-15},$$

$$\varepsilon_{32}^{(2)} \approx 8.4 \times 10^{-15},$$

$$\varepsilon_{32}^{(3)} \approx 4.7 \times 10^{-14}.$$

The Matlab code used to generate these results and Fig. 8.7 is provided. In order to go beyond the standard double precision accuracy, one has to use a specialized toolbox in multi-precision computations such as Advanpix [135].

```
1   l    = 1.0;       % half-length of the domain
2   N    = 128;       % number of Fourier modes
3   dx   = 2*l/N;     % distance between two collocation points
4   x    = (1-N/2:N/2)*dx; % physical space discretization
5
6   dk   = pi/l;
7   k    = [0:N/2 1-N/2:-1]*dk; % vector of wavenumbers
8
9   arg  = pi*(x + 1);
10  sar  = sin(arg);
11  car  = cos(arg); car2 = car.*car;
12  esa  = exp(sar);
13  u    = sar.*esa;
14  uhat = fft(u);
15
16  % numerical derivatives:
17  up   = real(ifft(1i*k.*uhat));
18  upp  = real(ifft(-k.^2.*uhat));
19  uppp = real(ifft(-1i*k.^3.*uhat));
20
21  % exact derivatives:
22  pi2  = pi*pi; pi3 = pi2*pi;
23  u1   = pi*car.*(1 + sar).*esa;
24  u2   = pi2*(car2.*(sar + 3) - 1 - sar).*esa;
25  u3   = pi3*car.*(car2.*(sar + 6) - 4 - 7*sar).*esa;
```

Above we exploit the property that the Fourier transform is a 'change a variables' where the differentiation operator ∂_x becomes diagonal.

8.4   Indications for Further Reading

If you got interested in the beautiful topic of pseudo-spectral methods, you can find more information in the following books:

- The following book by N. Trefethen[32] has two major advantages: (*i*) conciseness and (*ii*) collection of Matlab programs which comes along. We would recommend

---

[32]Lloyd Nick Trefethen (1955–20..), an American/British numerical analyst.

it as the first reading on pseudo-spectral methods. At least you will learn how to program efficiently in Matlab (or in Octave, Scilab, etc.) [189]:

- Trefethen [189] *Spectral methods in MatLab*. Society for Industrial and Applied Mathematics, Philadelphia, PA, USA.

- The book by J. Boyd[33] is probably the most exhaustive one. It covers many topics and applications of spectral methods. The material is presented as a collection of tricks. For example, it is one of seldom books where (semi-)infinite domains and spherical geometries are covered. We are not sure that after reading this book you will know how to program the pseudo-spectral methods, but you will have a broad of view of possible issues and how to address them [22]:

- Boyd [22] *Chebyshev and Fourier Spectral Methods*. (Dover Publications, New York) (2nd Ed.).

- This book is probably our favourite one. It represents a good balance between the theory and practice and this book arose as an extended version of a previously published review paper. The present lecture notes were inspired in part on this book as well [65]:

- Fornberg,[34] [65] *A practical guide to pseudospectral methods*. Cambridge: Cambridge University Press.

- The Authors discovered this book only recently. So, we have still to read it, but from the first sight I can already recommend it:

- Peyret [160]. *Spectral Methods for Incompressible Viscous Flow*. Springer-Verlag New York Inc.

- Finally, I discovered also these very clear and instructive Lecture notes [191]. Lectures were delivered in 2009 at the International Summer School '*Modern Computational Science*' in Oldenburg, Germany. They can be freely downloaded at the following URL address:

- http://www.staff.uni-oldenburg.de/hannes.uecker/pre/030-mcs-hu.pdf

In general, the Authors's suggestions (according to his personal taste and vision) for scientific literature and software are collected in a single document, which is continuously expanded and completed:

https://github.com/dutykh/libs/

---

[33]John Boyd (19..–20..), an American numerical analyst, meteorologist and occasional science fiction writer.

[34]Bengt Fornberg (19..–20..), a Swedish/American numerical analyst.

8.5   Appendix 1: Some Identities Involving Tchebyshev Polynomials

Tchebyshev polynomials $\{T_n(x)\}_{n=0}^{\infty}$ form a weighted orthogonal system on the segment $[-1, 1]$:

$$\langle T_n, T_m \rangle \equiv \int_{-1}^{1} \frac{T_n(x) \cdot T_m(x)}{\sqrt{1 - x^2}}\, dx = \begin{cases} 0, & m \neq n, \\ \pi, & m = n = 0, \\ \frac{\pi}{2}, & m = n > 0. \end{cases}$$

The first few Tchebyshev polynomials are

$$\begin{aligned} T_0(x) &= 1, \\ T_1(x) &= x, \\ T_2(x) &= 2x^2 - 1, \\ T_3(x) &= 4x^3 - 3x, \\ T_4(x) &= 8x^4 - 8x^2 + 1, \\ T_5(x) &= 16x^5 - 20x^3 + 5x. \end{aligned}$$

They are represented graphically in Fig. 8.8. Higher order Tchebyshev polynomials can be constructed using the three-term recursion relation:

$$T_{n+1}(x) = 2x\, T_n(x) - T_{n-1}(x).$$

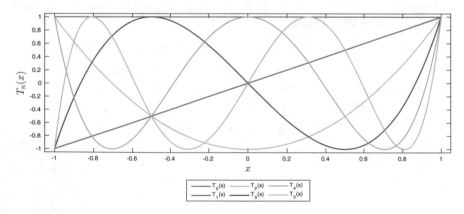

Fig. 8.8 The first five Tchebyshev polynomials. Notice that $T_n(-1) = (-1)^n$ and $T_n(1) \equiv 1$

The following Matlab code realizes this idea in practice:

```
1  function P = Chebyshev(n)
2      P = cell(n,1);
3      if (n == 1)
4          P{1} = 1;
5          return;
6      end % if ()
7
8      P{1} = 1;
9      P{2} = [1; 0];
10     for j=3:n
11         P{j}       = [2*P{j-1}; 0];
12         P{j}(3:j) = P{j}(3:j) - P{j-2};
13     end % for j
14
15 end % Chebyshev()
```

Then, nth Tchebyshev polynomial can be evaluated using the standard Matlab's function `polyval()`, e.g.

```
1  n  = 5;
2  P  = Chebyshev(n+1);
3  x  = linspace(-1, 1, 1000);
4  Tn = polyval(P{n+1}, x);
```

There is an explicit expression for the nth polynomial:

$$T_n(x) = \cos\big(n\theta(x)\big), \qquad \theta(x) \stackrel{\text{def}}{:=} \arccos x. \qquad (8.17)$$

The first derivatives of Tchebyshev polynomials can be constructed recursively as well:

$$\frac{T'_{n+1}(x)}{n + 1} = \frac{T'_{n-1}(x)}{n - 1} + 2\,T_n(x). \qquad (8.18)$$

Otherwise, the first derivative $T'_n(x)$ can be found from the following relation:

$$(1 - x^2)\,T'_n(x) = -n\,x\,T_n(x) + n\,T_{n-1}(x).$$

Tchebyshev polynomial $T_n(x)$ satisfies the following linear second order differential equation with non-constant coefficients:

$$(1 - x^2)\,T''_n - x\,T'_n + n^2\,T_n = 0.$$

Finally, zeros of the nth Tchebyshev polynomial $T_n(x)$ are located at

$$x_k^0 = \cos\left(\frac{2k - 1}{2n}\,\pi\right), \qquad k = 1, 2, \ldots, n,$$

and extrema at

$$x_k^{\text{ext}} = \cos\left(\frac{\pi k}{n}\right), \qquad k = 0, 1, \ldots, n.$$

When implementing Tchebyshev-type pseudo-spectral methods for nonlinear problems, one can use the expression of the product of two Tchebyshev polynomials in terms of higher and lower degree polynomials:

$$T_m(x)\, T_n(x) = \frac{1}{2}\left(T_{n+m}(x) + T_{n-m}(x)\right), \qquad \forall n \geq m \geq 0.$$

To finish this Appendix we give the *generating function* for Tchebyshev polynomials:

$$\frac{1 - xz}{1 - 2xz + z^2} = \sum_{n=0}^{\infty} T_n(x)\, z^n.$$

In order to satisfy the requirement (2) from Sect. 8.1.1, we provide here the relations which allow to re-express the derivatives of the Tchebyshev expansion in terms of Tchebyshev polynomials again. Namely, consider a truncated series expansion of a function $u(x)$ in Tchebyshev polynomials:

$$u(x) = \sum_{k=0}^{N} v_k\, T_k(x).$$

Imagine that we want to compute the first derivative of the expansion above (it is always possible, since the sum is finite and we can differentiate term by term):

$$u'(x) = \sum_{k=0}^{N} v_k\, \frac{dT_k(x)}{dx}.$$

Now we re-expand the derivative $u'(x)$ in the same basis functions:

$$u'(x) = \sum_{k=0}^{N} v_k'\, T_k(x).$$

The main question is how to re-express the coefficients $\{v_k'\}_{k=0}^{N}$ in terms of coefficients $\{v_k\}_{k=0}^{N}$? This goal is achieved using basically the recurrence relation (8.18) (even if it does not jump into the eyes). Thanks to it we have an explicit relation between the coefficients:

$$v_k' = \frac{2}{\delta_k} \sum_{\substack{j=k+1 \\ j+k \ \text{odd}}}^{N} j\, v_j, \qquad j = 0, 1, \ldots, N-1,$$

where

$$\delta_k = \begin{cases} 2, & k = 0, \\ 0, & k \neq 0. \end{cases}$$

A similar 'trick' can be made for the 2nd derivative as well:

$$u''(x) = \sum_{k=0}^{N} v_k \frac{d^2 T_k(x)}{dx^2} = \sum_{k=0}^{N} v_k'' T_k(x).$$

The connection between coefficients $\{v_k''\}_{k=0}^{N}$ and $\{v_k\}_{k=0}^{N}$ is given by the following explicit formula:

$$v_k'' = \frac{1}{\delta_k} \sum_{\substack{j=k+2 \\ j+k \text{ even}}}^{N} j \, (j^2 - k^2) v_j, \qquad j = 0, 1, \ldots, N-2.$$

Finally, in order to construct the spectral coefficients for the nth derivative, one can use the following recurrence relation (also stemming from (8.18)):

$$\delta_{k-1} v_{k-1}^{(n)} = v_{k+1}^{(n)} + 2k v_k^{(n-1)}, \qquad k \geq 1,$$

which has to be completed by starting value $v_N' \equiv 0$.

Remark 8.8 Similar identities exist for the family of Jacobi polynomials $\mathcal{P}_n^{\alpha, \beta}(x)$ as well. However, they are more complicated due to the presence of two arbitrary parameters α, $\beta > -1$. The family of Tchebyshev polynomials is a particular case of Jacobi polynomials when $\alpha = \beta = -\frac{1}{2}$.

8.5.1 Compositions of Tchebyshev Polynomials

We would like to show here another interesting formula involving Tchebyshev polynomials:

Theorem 8.2 (Composition formula) *If $T_n(x)$ and $T_m(x)$ are Tchebyshev polynomials ($m, n \geq 0$), then*

$$\left(T_m \circ T_n\right)(x) \equiv T_m\big(T_n(x)\big) = T_{mn}(x). \tag{8.19}$$

Proof An *eleven* (!) pages combinatorial proof of this result can be found in [13]. Here we shall prove the composition formula (8.19) in one line by using some rudiments of complex variables. Let us introduce the variable $z \overset{\text{def}}{:=} e^{i\theta} \in \circlearrowright^{35}$ such

---

[35]Symbol $\circlearrowright$ denotes the unit circle $\mathscr{S}_1(0)$ on the complex plane $\mathbb{C}$.

that

$$x = \frac{1}{2}\left(z + \frac{1}{z}\right) \quad \leftrightarrows \quad x = \cos\theta.$$

Then, by (8.17) we have a complex representation of the nth Tchebyshev polynomial:

$$T_n(x) \equiv T_n\left[\tfrac{1}{2}\left(z + z^{-1}\right)\right] = \frac{1}{2}\left(z^n + \frac{1}{z^n}\right).$$

Now, we have all the elements to show the main result:

$$T_m\left[T_n\left(\tfrac{1}{2}\left(z + z^{-1}\right)\right)\right] = T_m\left[\tfrac{1}{2}\left(z^n + z^{-n}\right)\right] = \frac{1}{2}\left((z^n)^m + \frac{1}{(z^n)^m}\right) =$$

$$\frac{1}{2}\left(z^{mn} + \frac{1}{z^{mn}}\right) = T_{mn}\left[\tfrac{1}{2}\left(z + z^{-1}\right)\right] \equiv T_{mn}(x).$$

8.6   Appendix 2: Trefftz Method

In this Appendix we would like to give the flavour of the so-called Trefftz[36] methods, which remain essentially unknown/forgotten nowadays. These methods belong
to boundary-type solution procedures. Below I shall quote a Professor[37] from Laboratoire Jacques-Louis Lions (LJLL) at Paris 6 Pierre and Marie Curie University,
who makes interesting comments about the knowledge of Trefftz methods in French
Applied Mathematics community (the original language and orthography are conserved):

> […] Sinon j'ai moi aussi fait une petite enquête, à la suite de laquelle il apparaît que personne
> ne connait Trefftz chez les matheux de Paris 6, hormis Nataf qui en a entendu parler pendant
> son DEA en méca!!!
>
> Trefftz est un grand oublié car son papier est vraiment extrêmement intéressant, encore plus
> si tu te rends compte qu'il produit en fait une estimation *a posteriori* (ça doit même être la
> première). En revanche je me suis aussi persuadé que les méthodes de Trefftz végètent chez
> les mécano. […]

So, the situation is rather sad. Fortunately this method continued to live in the
Mechanical Engineering community who preserved it for future generations. In the
exposition below we shall follow an excellent review paper [115]. More precisely
we describe the so-called *indirect Trefftz method*, which was proposed originally by
Trefftz (1926) in [190]. There are also *direct Trefftz methods* proposed some sixty
years later in [37]. They are much closer to the Boundary Integral Equation Methods
(BIEM) and will not be covered here. Interested readers can refer to [115].

---

[36]Erich Emmanuel Trefftz (1888–1937), a German Mathematician and Mechanical Engineer.

[37]His/Her identity will be hidden in order to avoid any kind of diplomatic incidents.

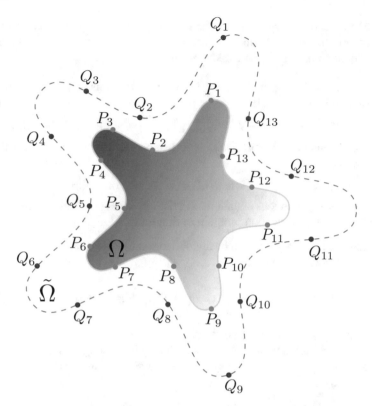

Fig. 8.9 Collocation Trefftz method using the Green function $\mathscr{G}(P, Q)$: $\{P_i\}_{i=1}^{13} \subseteq \mathbb{R}^d$ are collocation points and $\{Q_i\}_{i=1}^{13} \subseteq \mathbb{R}^d$ arc the sources put outside of the domain Ω in order to avoid singularities. Then, the approximate solution is sought as a linear combination of functions $\{\mathscr{G}(P, Q_i)\}_{i=1}^{13}$. The unknown expansion coefficients are found so that to satisfy the boundary conditions exactly in collocation points $\{P_i\}_{i=1}^{13}$. The outer points $\{Q_i\}_{i=1}^{13}$ are not needed if we know a $\mathbb{T}$-complete set of functions for the operator $\mathscr{L}$

Consider a (compact) domain $\Omega \subseteq \mathbb{R}^d, d \geq 2$ and a (scalar) equation on it (see Fig. 8.9 for an illustration):

$$\mathscr{L}u = 0. \tag{8.20}$$

You can think, for example, that $\mathscr{L} = -\nabla^2 = -\sum_{i=1}^d \dfrac{\partial^2}{\partial x_i^2}$ is the Laplace[38] operator. Equation (8.20) has to be completed by appropriate boundary conditions:

$$\mathscr{B}u = u^\circ, \quad x \in \partial\Omega, \tag{8.21}$$

---

[38]Pierre–Simon de Laplace (1749–1827) is a French Mathematician whose works were greatly disregarded during his times. The understanding of their importance came much later (il vaut mieux tard que jamais ☺).

where u° is the boundary data (solution value or its flux) and $\mathscr{B}$ is an operator depending on the type of boundary conditions in use. For example, in the case of Laplace equation the following choices are popular:

Dirichlet $\quad \mathscr{B} \overset{\text{def}}{:=} \mathbb{I}$ (Identity operator, i.e. $\mathbb{I}u \equiv u$); in this case the solution value is prescribed on the boundary.

Neumann $\quad \mathscr{B} \overset{\text{def}}{:=} \dfrac{\partial}{\partial n} \equiv \mathbf{n} \cdot \nabla = \sum_{i=1}^{d} n_i \cdot \dfrac{\partial}{\partial x_i}$ (Normal derivative, $\mathbf{n}$ being the exterior normal to $\partial \Omega$); physically it corresponds to the prescribed flux through the boundary.

Robin $\quad \mathscr{B} \overset{\text{def}}{:=} \alpha \mathbb{I} + \beta \dfrac{\partial}{\partial n}$, where $\alpha, \beta \in \mathbb{R}$ are some parameters; this case is a mixture of two previous situations. It arises in some problems as well (see e.g. [38] for the heat conduction problem in thin liquid films).

Notice, that the boundary $\partial \Omega$ can be divided in some sub-domains where a different boundary condition is imposed:

$$\partial \Omega = \Gamma_1 \cup \Gamma_2 \cup \ldots \cup \Gamma_n \qquad \mu_d \left(\Gamma_i \cap \Gamma_j \right) = 0, \quad 1 \leq i < j \leq n,$$

where μ_d is the Lebesgue measure in $\mathbb{R}^d$.

In the indirect Trefftz method the numerical solution is sought as a linear combination of $\mathbb{T}$-complete functions $\{\phi_k(\mathbf{x})\}_{k=1}^{n}$ [97], which satisfy exactly the governing equation (8.20):

$$u_n(\mathbf{x}) = \sum_{k=1}^{n} v_k \, \phi_k(\mathbf{x}) . \tag{8.22}$$

For example, for the Laplace equation in $\mathbb{R}^2$ the $\mathbb{T}$-complete set of functions is [96] $\{r^k e^{ik\theta}\}_{k=0}^{\infty}$, where (r, θ) are polar coordinates[39] on plane. The coefficients $\{v_k\}_{k=1}^{n}$ are to be determined. To achieve this goal the approximate solution (8.22) is substituted into the boundary conditions (8.21) to form the residual:

$$\mathscr{R}[u_n] = \mathscr{B} u_n - u^\circ .$$

If the residual $\mathscr{R}[u_n]$ is equal identically to zero on the boundary $\partial \Omega$, then we found the exact solution (we are lucky!). Otherwise (we are unlucky and in Numerical Analysis (NA) it happens more often), the residual has to be minimized. Very often the boundary operator $\mathscr{B}$ is linear and we can write:

$$\mathscr{R}[u_n] = \sum_{k=1}^{n} v_k \, \mathscr{B} \phi_k(\mathbf{x}) - u^\circ .$$

---

[39]$r = \sqrt{x^2 + y^2}, \quad \theta = \arctan \dfrac{y}{x}, \quad$ with $\arctan(\pm\infty) = \pm\dfrac{\pi}{2}$.

So, we can apply now the collocation, Galerkin or least square methods to determine the coefficients $\{v_k\}_{k=1}^n$. These procedures were explained above.

Remark 8.9 Suppose that a $\mathbb{T}$-complete set of functions for Eq. (8.20) is unknown. However, there is a way to overcome this difficulty if we know the Green[40] function[41] $\mathscr{G}(x; Q), x \in \Omega$ for our Eq. (8.20), i.e.

$$\mathscr{L}\mathscr{G} = \delta(x - Q), \quad x \in \Omega, \quad Q \in \mathbb{R}^d.$$

where $\delta(x)$ is the singular Dirac measure. Notice that point Q can be inside or outside of domain Ω, where the problem is defined. Then, we can seek for an approximate solution $u_n(x)$ as a linear combination of Green functions:

$$u_n(x) = \sum_{k=1}^n v_k \mathscr{G}_k(x; Q_k),$$

where the points $\{Q_k\}_{k=1}^n$ are distributed outside of the computational domain Ω in order to avoid the singularities. This method is schematically illustrated in Fig. 8.9. The accuracy of this approach depends naturally on the distribution of points $\{Q_k\}_{k=1}^n$. To Author's knowledge there exist no theoretical indications for the optimal distribution (as it is the case of Tchebyshev nodes on a segment). So, it remains mainly an experimental area of the research.

Conclusions

In the indirect Trefftz methods (historically, the original one) the problem solution is sought as a linear combination of the functions, which satisfy the governing equation identically. Then, the unknown coefficients are chosen so that the approximate solution satisfies the boundary condition(s) as well. It can be done by means of collocation, Galerkin or least square procedures. The main advantage of Trefftz methods is that it allows to reduce problem's dimension by one (i.e. 3D ⤳ 2D and 2D ⤳ 1D), since the residual $\mathscr{R}[u_n]$ is minimized on the domain boundary $\partial\Omega \subseteq \mathbb{R}^{d-1}$. Readers who are interested in the application of Trefftz methods to their problems should refer to [115] and references therein.

---

[40]George Green (1793–1841), a British Mathematician who made great contributions to the Mathematical Physics and PDEs.

[41]The Green function is named after George Green (see the footnote above), but the notion of this function can be already found first in the works of Laplace, then in the works of Poisson. As I said, the works of Laplace were essentially disregarded by his "colleagues". Poisson was luckier in this respect. He is also known for (mis)using his administrative resource in order to delay (just for a couple of decades, nothing serious ☺) the publication of his competitors, e.g. the young (at that time) Cauchy.

8.7   Appendix 3: Monte–Carlo Approach to the Diffusion Simulation

From considering the history and physical modelling of Brownian motion we naturally come to the numerical simulation of parabolic PDEs using stochastic processes. It falls into the large class of Monte–Carlo and quasi-Monte–Carlo methods [31]. They are based on the so-called *Law of large numbers*, which can be informally stated[42] as

$$\lim_{N \to +\infty} \frac{1}{N} \sum_{n=1}^{N} g(\xi_n) = \mathrm{e}\,[g(\xi)],$$

where ξ_n are independent, identically distributed random variables and $g(\cdot)$ is a real-valued continuous function. Here $\mathrm{e}\,[\cdot]$ stands for the mathematical expectation and the convergence is understood in the sense *convergence in probability distribution*.[43] The proof is based on the Tchebyshev inequality in Probabilities, but we do not enter into these details. The main advantages of Monte–Carlo methods are

- Simplicity of implementation
- Independent of the problem dimension (they do not suffer of the curse of dimensionality)

On the other hand, the convergence rate is given by the Central Limit Theorem [104] and it is rather slow, i.e. $\mathscr{O}(N^{-1/2})$, even if some acceleration is possible thanks to some adaptive procedures [121] (such as low-discrepancy sequences, variance reduction and multi-level methods). The large deviation theory guarantees that the probability of falling out of a fixed tolerance interval decays exponentially fast.

To the Author knowledge, the method we are going to describe below was first proposed by R. Feynman[44] in order to solve *numerically* the linear Schrödinger[45] equation in 1940s. In fact, he noticed that the Schrödinger equation can be solved by a kind of average over trajectories. This observation led him to a far-reaching reformulation of the quantum theory in terms of *path integrals*. Upon learning Feynman's ideas, M. Kac[46] (Feynman's colleague at Cornell University) understood that a similar method can work for the heat equation as well. Later it became Feynman–Kac

---

[42]We choose this particular form, since it is suitable for our exposition.

[43]A sequence of random variables $\{\xi_n\}_{n=1}^{\infty}$ *converges in probability distribution* towards the random variable ξ if for $\forall \varepsilon > 0$

$$\lim_{n \to \infty} \mathbb{P}\{|\xi_n - \xi| \geq \varepsilon\} = 0.$$

This fact can be denoted as $\xi_n \overset{\mathbb{P}}{\rightsquigarrow} \xi$.

[44]Richard Phillips Feynman (1918–1988) was an American theoretical Physicist. Nobel Prize in Physics (1965) and Author's hero. Please, do not hesitate to read any of his books!

[45]Erwin Rudolf Josef Alexander Schrödinger (1887–1961) was an Austrian theoretical Physicist. Nobel prize in Physics (1933) for the formulation of what is known now as the Schrödinger equation.

[46]Mark Kac (1914–1984) was a Polish mathematician who studied in Lviv University, Ukraine and immigrated later to USA.

method. Unfortunately, this method is not implemented in any commercial software (for PDEs, the financial industry is using these methods since many decades) and it is not described in classical textbooks on PDEs and in the books on numerical methods for PDEs.

Consider first the following simple heat equation:

$$u_t = \frac{1}{2} u_{xx} - V(x) \cdot u, \tag{8.23}$$

where $V(x)$ is a function representing the amount of external cooling (if $V(x) \geq 0$, and external heating if $V(x) < 0$, but we do not consider this case) at point x. Then, we have the following

Theorem 8.3 (Feynman–Kac formula) *Let $V(x)$ be a non-negative continuous function and let $u_0(x)$ be bounded and continuous. Suppose that $u(x, t)$ is a bounded function that satisfies Eq. (8.23) along with the initial condition*

$$u(x, 0) = u_0(x), \tag{8.24}$$

then,

$$u(x, t) = e\left[\exp\left\{ -\int_0^t V(\mathscr{W}_s) \, ds \right\} u_0(\mathscr{W}_t) \right], \tag{8.25}$$

where $\{\mathscr{W}_t\}_{t \geq 0}$ is a Brownian motion starting at x.

We have to define rigorously also what the Brownian motion is:

Definition 1 (*Brownian motion*) A real-valued stochastic process $t \mapsto \mathscr{W}(t)$ (or a random curve) is called a *Brownian motion* (or *Wiener process*) if

1. Almost surely $\mathscr{W}(0) = 0$
2. $\mathscr{W}(t) - \mathscr{W}(s) \sim \mathscr{N}(0, t - s) \equiv \sqrt{t - s} \, \mathscr{N}(0, 1), \quad \forall t \geq s \geq 0$
3. For any instances of time $v > u > t > s \geq 0$ the increments $\mathscr{W}(v) - \mathscr{W}(u)$ and $\mathscr{W}(t) - \mathscr{W}(s)$ are independent random variables.

In particular, from point (2) by choosing $s = 0$ it follows directly that

$$e[\mathscr{W}(t)] = 0, \qquad e[\mathscr{W}^2(t)] = t.$$

More informally we can say that the Brownian motion is a continuous random curve with the *largest* possible amount of randomness.

The Theorem above can be proved even under more general assumptions, but the formulation given above suffices for most practical situations. For instance, the functions $V(x)$ and $u_0(x)$ may have isolated discontinuities and the Feynman–Kac formula will be still valid. Another interesting corollary of Feynman–Kac's formula is the uniqueness of solutions to the Initial Value Problem (IVP) (8.23):

Corollary 8.1 *Under the assumptions of Theorem 8.3, there is at most one solution to the heat equation (8.23), which satisfies the initial condition (8.24). Namely, it is given by Feynman–Kac formula (8.25).*

The main (computational) advantage of Feynman–Kac formula is that it can be straightforwardly generalized to the arbitrary dimension:

Theorem 8.4 *(Feynman–Kac formula in d dimensions) Let $V : \mathbb{R}^d \mapsto [0, +\infty)$ and $u_0 : \mathbb{R}^d \mapsto \mathbb{R}$ be continuous functions with bounded initial condition $u_0(x)$. Suppose that $u(x, t)$ is also a bounded function, which satisfies the following Partial Differential Equation (PDE):*

$$u_t = \frac{1}{2} \sum_{i=1}^{d} \frac{\partial^2 u}{\partial x_i^2} - V(x) \cdot u,$$

and the initial condition

$$u(x, 0) = u_0(x).$$

Then,

$$u(x, t) = e\left[\exp\left\{ -\int_0^t V(\mathscr{W}_s)\, ds \right\} u_0(\mathscr{W}_t) \right],$$

where $\{\mathscr{W}_t\}_{t \geq 0}$ is a d-dimensional Brownian motion starting at x.

We give here another useful generalization of the Feynman–Kac method. For the sake of simplicity we return to the one-dimensional case. Consider a stochastic process X_t, which satisfies the following *Stochastic Differential Equation* (SDE):

$$dX_t = \alpha(X_t)\, dt + \sigma(X_t)\, d\mathscr{W}_t, \qquad X_0 = x, \tag{8.26}$$

where $\alpha(x)$ is the local drift and $\sigma(x)$ is called the *local volatility* in financial applications. We assume here that $\alpha(x)$ and $\sigma(x)$ are globally Lipschitz continuous and grow linearly in space at most. Then, the function given by formula (8.25) satisfies the generalized diffusion equation

$$u_t = \alpha(x)\, u_x + \frac{1}{2} \sigma^2(x)\, u_{xx} - V(x)\, u(x),$$

together with the initial condition (8.24).

So, the resulting numerical algorithm is very simple:

1. Generate N trajectories of the Brownian motion $\{\mathscr{W}_s^k\}_{s \in [0, t]}^{1 \leq k \leq N}$
2. Compute N solutions to the SDE (8.26) using the Euler–Maruyama[47] method [99], for example

---

[47]Gisiro Maruyama (1916–1986) was a Japanese Mathematician with notable contributions to the theory of stochastic processes.

3. Compute the solution value using representation (8.25). The mathematical expectation e [·] is replaced by the simple arithmetic average according to the Monte–Carlo approach.

Remark 8.10 Notice that the step (2) can be omitted if we solve the linear parabolic equation without the drift term and with constant diffusion coefficient. The simplest version of the Feynman–Kac method for the simplest initial value problem

$$u_t = \frac{1}{2} u_{xx}, \qquad u(x, 0) = u_0(x),$$

looks like

$$u(x, t) \approx \frac{1}{N} \sum_{n=1}^{N} u_0(\xi_n), \qquad \forall n : \xi_n \sim \mathcal{N}(x, \sqrt{t}).$$

It is probably the simplest way to estimate solution value in a given point (x, t).

The Feynman–Kac method inherits all advantages (and disadvantages) of Monte–Carlo methods. Comparing to grid-based methods it enjoys also the locality property. Namely, if you want to compute your solution in a given point (x, t), you do not need to know the solution in neighbouring sites (locations). So, if you want to compute numerically the solution only in a few specific points, please, consider this method. It can be competitive with more conventional approaches. Moreover, the financial industry already appreciated the power of these methods.

8.7.1 Brownian Motion Generation

In this section we provide a simple Matlab code to generate M sample Brownian motion realizations. It can be used as a building block for SDE solvers and practical implementations of the Feynman–Kac method described above:

```
1   M   =  100;   % number of paths
2   N   =  1000;  % number of steps
3   T   =  1;     % final simulation time
4   dt  =  T/N;   % time step
5   dW  =  sqrt(dt)*randn(M, N);
6   W   =  cumsum(dW, 2);
```

A sample output result (with precisely the same parameters) of this script is depicted in Fig. 8.10.

It is quite straightforward to employ the same Matlab script to generate Brownian paths in d dimensions (just take $M = m \times d$, where $m \in \mathbb{N}$ and then regroup matrix W in m d-dimensional trajectories). A few realizations of the Brownian motion in two spatial dimensions are depicted in Fig. 8.11.

Fig. 8.10 A collection of sample Brownian paths generated by the code given in Appendix 8.7.1. Two standard deviations should contain about 99% of trajectories. On this picture it looks like it is the case

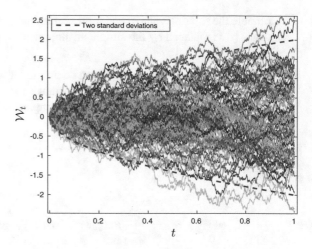

Fig. 8.11 A few Brownian paths in two spatial dimensions. The red circle depicts the starting point $(0, 0)$

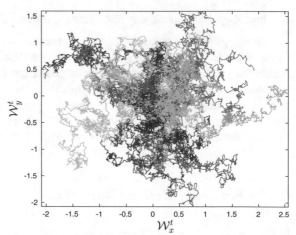

8.8   Appendix 4: An Exact Non-periodic Solution to the 1D Heat Equation

In this Appendix we provide an exact solution (see [65, Sect. §7.2]) to the (dimensionless) heat equation

$$u_t - \frac{1}{9\pi^2}\, u_{xx} = 0\,, \quad x \in [0,\, 1]\,, \quad t \in \mathbb{R}^+\,, \tag{8.27}$$

subject to the initial condition

$$u(x,\, 0) = 0\,, \quad x \in [0,\, 1]\,,$$

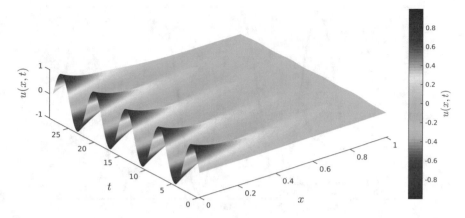

Fig. 8.12 Space-time plot of the solution (8.28) to the linear heat equation (8.27). The time window is $t \in [0, 9\pi]$

and the following non-periodic and non-homogeneous boundary conditions:

$$u(0, t) = \sin t, \qquad u_x(1, t) = 0, \qquad t \in \mathbb{R}^+.$$

The first boundary condition says that the temperature is prescribed at the left boundary and zero heat flux is imposed on the right boundary. So, the unique solution to the problem (8.27) described above for $\forall t > 0$ is given by[48]

$$u(x, t) = \underbrace{\sinh^{-1} \frac{3\pi}{2} \left[\cos \frac{3\pi x}{2} \sinh \frac{3\pi(1 - x)}{2} \sin t - \sin \frac{3\pi x}{2} \cosh \frac{3\pi(1 - x)}{2} \cos t \right]}_{u_\circlearrowleft(x, t)}$$

$$+ \underbrace{\frac{72}{\pi} \sum_{n=1}^{\infty} \frac{(2n - 1) e^{-\frac{(2n-1)^2}{18} t}}{\left[9 + 4(n - 2)^2\right]\left[9 + 4(n + 1)^2\right]} \sin(n - \tfrac{1}{2})\pi x}_{u_\Sigma(x, t)} .$$

$$(8.28)$$

This analytical solution can be used as the *reference solution* in order to validate your non-periodic numerical codes (see Fig. 8.12 for an illustration). Notice that the second term $u_\Sigma(x, t)$ (i.e. the infinite series) vanishes uniformly in space, i.e.

$$u_\Sigma(x, t) \rightrightarrows 0 \quad \text{as} \quad t \to +\infty.$$

This term u_Σ is needed to enforce the initial condition. Consequently, the long time behaviour of the solution (8.28) is given by $u_\circlearrowleft(x, t)$.

---

[48]This solution can be derived using the classical method of separation of variables [84] [or see an excellent publication of Professor (and my favourite colleague and friend) Marguerite Gisclon [80], if you are not scared by French].

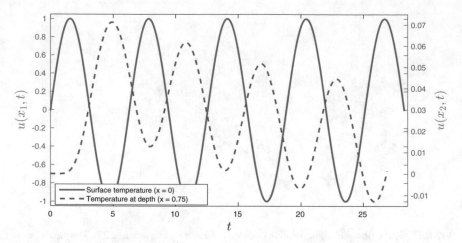

Fig. 8.13 Solution (8.28) shown at the surface $x = 0$ and at depth $x = \frac{3}{4}$. The goal is to illustrate the phase shift between two curves

Remark 8.11 Equation (8.27) (along with boundary conditions) models the temperature variation in soil under periodic boundary forcing (modeling seasonal temperature variations). Solution (8.28) explains also that for many types of soils, there is a phase shift of seasonal temperature at certain depths (warmest in winter and coolest in summer). See Fig. 8.13 for an illustration. It provides a theoretical explanation why a cave works in practice.

8.9   Some Popular Numerical Schemes for ODEs

The discussion of numerical methods for PDEs cannot be complete if we do not provide some basic techniques for the time marching. Basically, if we follow the Method Of Lines (MOL) [117, 166, 177, 180], the PDE is discretized first in space, then the resulting system of coupled ODEs has to be solved numerically. This document would not be complete if we did not provide any indications on how to do it. Consider for simplicity an Initial Value Problem (IVP), which is sometimes also referred to as the Cauchy problem:

$$\dot{u} = f(u), \qquad u(0) = u_0.$$

where the dot over a function denotes the derivative with respect to time, i.e. $\dot{u} \equiv \dfrac{du}{dt}$. There is a number of time marching schemes proposed in the literature. We refer to [92–94] as exhaustive references on this topic. Below we provide some most popular (subjectively) schemes.

Forward Euler (explicit, first order accurate)

$$u_{n+1} = u_n + \Delta t \, f(u_n).$$

Backward Euler (implicit, first order accurate)

$$u_{n+1} = u_n + \Delta t \, f(u_{n+1}).$$

Adams[49]–Bashforth-2 [49] (explicit, second order accurate)

$$u_{n+1} = u_n + \Delta t \left[\frac{3}{2} f(u_n) - \frac{1}{2} f(u_{n-1}) \right].$$

Adams–Bashforth[50]-3 [50] (explicit, third order accurate)

$$u_{n+1} = u_n + \Delta t \left[\frac{23}{12} f(u_n) - \frac{4}{3} f(u_{n-1}) + \frac{5}{12} f(u_{n-2}) \right].$$

Adams–Moulton[51]-1 or the trapezoidal rule [51] (implicit, second order accurate)

$$u_{n+1} = u_n + \frac{1}{2} \Delta t \left[f(u_{n+1}) + f(u_n) \right].$$

Adams–Moulton-2 (implicit, third order accurate)

$$u_{n+1} = u_n + \Delta t \left[\frac{5}{12} f(u_{n+1}) + \frac{2}{3} f(u_n) - \frac{1}{12} f(u_{n-1}) \right].$$

Two-stage Runge–Kutta[52]2 [52] (explicit, second order accurate)

$$k_1 = \Delta t \, f(u_n),$$
$$k_2 = \Delta t \, f(u_n + \alpha \, k_1),$$
$$u_{n+1} = u_n + \left(1 - \frac{1}{2\alpha} \right) k_1 + \frac{1}{2\alpha} k_2.$$

---

[49] John Couch Adams (1819–1892), a British Mathematician and Astronomer. He predicted the existence and position of the planet Neptune.

[50] Francis Bashforth (1819–1912) is a British Applied Mathematician working in the field of Ballistics. However, his famous numerical scheme was proposed in collaboration with J. C. Adams to study the drop formation.

[51] Forest Ray Moulton (1872–1952) is an American Astronomer. There is a crater on the Moon named after him.

[52] Martin Wilhelm Kutta (1867–1944) is a German Mathematician who worked in the field of Fluid Mechanics and Numerical Analysis.

- $\alpha = \frac{1}{2}$: mid-point scheme
- $\alpha = 1$: Heun's[53] method
- $\alpha = \frac{2}{3}$: Ralston's method

Runge–Kutta-4 (RK4)   (explicit, fourth order accurate)

$$k_1 = \Delta t \, f(u_n),$$

$$k_2 = \Delta t \, f\left(u_n + \frac{1}{2}\, k_1\right),$$

$$k_3 = \Delta t \, f\left(u_n + \frac{1}{2}\, k_2\right),$$

$$k_4 = \Delta t \, f(u_n + k_3),$$

$$u_{n+1} = u_n + \frac{1}{6}\left[k_1 + 2\,k_2 + 2\,k_3 + k_4\right].$$

We do not discuss here the questions related to the stability of these schemes. This topic is of uttermost importance but out of scope of this chapter. Moreover, we skip also the adaptive embedded Runge–Kutta schemes which can choose the time step to meet some prescribed accuracy requirements [53].

In order to illustrate the usage of RK4 scheme (sometimes called *the* Runge–Kutta scheme) we take a simple nonlinear logistic equation:

$$\dot{u} = u \cdot (1 - u), \quad u(0) = 2. \tag{8.29}$$

It is not difficult to check that the exact solution to this IVP is

$$u(t) = \frac{2}{2 - e^{-t}}.$$

The Matlab code used to study numerically the convergence of the RK4 scheme on the nonlinear Eq. (8.29) is given below:

```
1   T   = 2.0; % final simulation time
2   uex = 2/(2 - exp(-T));
3   rhs = @(u) u*(1 - u); % RHS of the logistic equation
4
5   % list of successfully refined grids:
6   NN  = [100; 150; 200; 250; 350; 500; 750; 900; 1000; 1250; 1500;
7          2000; 2500; 3000; 3500; 4000; 4500; 5000; 5500; 6000];
8
9   Err = zeros(size(NN));
10  for n = 1:length(NN)
11      N   = NN(n); % number of time steps
12      dt  = T/N;   % one time step
13      u   = 2.0;   % initial condition on the solution
14      for j = 1:N
```

---

[53] Karl Heun (1859–1929) is a German Mathematician.

```
15          k1 = dt*rhs(u);
16          k2 = dt*rhs(u + 0.5*k1);
17          k3 = dt*rhs(u + 0.5*k2);
18          k4 = dt*rhs(u + k3);
19          u  = u + (k1 + 2*k2 + 2*k3 + k4)/6;
20       end % for j
21       Err(n) = abs(u - uex);
22    end % for n
```

The numerical result is shown in Fig. 8.14. One can clearly observe the 4th order convergence of the RK4 scheme applied to a nonlinear example. The convergence is broken only by the rounding effects inherent to the floating point arithmetics.

8.9.1 Existence and Unicity of Solutions

Normally, before attempting to solve a differential equation, one has to be sure that the mathematical problem is well-posed. The mathematical notion of the well-posedness was proposed by Hadamard[54] [90] and it includes three points to be checked:

1. Existence
2. Uniqueness
3. Continuous dependence on the initial condition and other problem parameters

The last point is usually more difficult to be proven theoretically. However, for 3D Navier[55]–Stokes[56] equations even the global existence and smoothness of solutions (i.e. the first point) poses already a Millennium problem:

> http://www.claymath.org/millennium-problems/navier-stokes-equation

Below we give some theoretical results which address the well-posedness conditions for an Initial Value Problem (IVP) for a scalar equation:

$$\dot{u} = f(t, u), \qquad u(t_0) = u_0. \tag{8.30}$$

The generalization to systems of Ordinary Differential Equations (ODEs) is straightforward.

---

[54]Jacques Hadamard (1865–1963), a French Mathematician who made seminal contributions to several fields of Mathematics including the Number Theory, Complex Analysis and the theory of PDEs.

[55]Claude Louis Marie Henri Navier (1785–1836) is a French Engineer (Corps des Ponts et Chaussées) and Physicist who made contributions to Mechanics.

[56]George Gabriel Stokes (1819–1903) was an Irish Physicist and Mathematician.

Fig. 8.14 Numerical
convergence of the 4th order
Runge–Kutta scheme when
applied to the logistic
equation

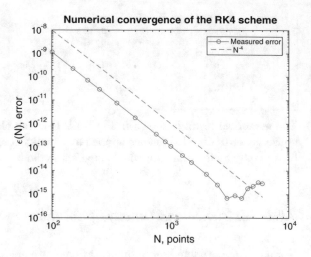

Theorem 8.5 (Existence) *If function $f(t, u)$ is continuous and bounded in a domain $(t, u) \in \mathcal{D} \subseteq \mathbb{R}_t \times \mathbb{R}_u$, then through every point $(t_0, u_0) \in \mathcal{D}$ passes at least one integral curve of Equation* (8.30).

Theorem 8.6 (Prolongation) *Let function $f(t, u)$ is defined and continuous in domain $\mathcal{D}$. If a solution $u = \phi(t)$ to problem* (8.30) *exists in the interval $t \in [t_0, \alpha)$ cannot be prolongated beyond the point $t = \alpha$, then it can happen only for one of three following reasons:*

1. $\alpha = +\infty$,
2. *When $t \rightarrow \alpha - 0$, $|\phi(t)| \rightarrow +\infty$,*
3. *When $t \rightarrow \alpha - 0$, distance between the point $\left(t, \phi(t)\right)$ to the boundary $\partial \mathcal{D}$ goes to zero.*

Theorem 8.7 (Uniqueness, [149]) *If function $f(t, u)$ is defined in domain $\mathcal{D}$ and for any pair of points (t_1, u_1), $(t_2, u_2) \in \mathcal{D}$ satisfies the condition*

$$|f(t_1, u_1) - f(t_2, u_2)| \leq \omega\left(|u_1 - u_2|\right), \qquad (8.31)$$

where $\omega(u) > 0$ is continuous for $0 < u \leq \mathcal{U}$ and

$$\int_{t_0 + \varepsilon}^{\mathcal{U}} \frac{du}{\omega(u)} \rightarrow +\infty, \qquad as \qquad \varepsilon \rightarrow 0,$$

then through any point $(t_0, u_0) \in \mathcal{D}$ passes at most one integral curve of Eq. (8.30).

Suitable functions $\omega(u)$ which satisfy the conditions of the Theorem are

$$\omega(u) = \mathcal{K} u \,,$$
$$\omega(u) = \mathcal{K} u \, |\ln u| \,,$$
$$\omega(u) = \mathcal{K} u \, |\ln u| \cdot \ln |\ln u| \,,$$
$$\omega(u) = \mathcal{K} u \, |\ln u| \cdot \ln |\ln u| \cdot \ln\ln |\ln u| \,, \qquad \text{etc.}$$

Above $\mathcal{K}$ is a positive constant. If we take $\omega(u) = \mathcal{K} u$ then condition (8.31) becomes the well-known Lipschitz[57] condition for function $f(t, u)$ in the second variable u. In order to satisfy the Lipschitz condition, it is sufficient for function $f(t, u)$ to be defined in a domain $\mathscr{D}$ convex in u and to have a bounded derivative $\dfrac{\partial f}{\partial u}$ in this domain. Later, Wintner showed that Osgood[58]'s Theorem conditions are sufficient for the convergence of Picard[59]'s iterations to a local solution on a sufficiently small interval [199].

8.9.1.1 Counterexamples

The Theorem above gives the existence of solutions only *locally* in time $[t_0, t_0 + \delta)$. The only reason for a solution not to exist beyond the given interval is the *blow-up* (i.e. the solution becomes unbounded). Otherwise, the solution exists globally. For instance, the following problem

$$\dot{u} = 1 + u^2, \qquad u(0) = 0 \,,$$

has the unique exact solution $u(t) = \tan(t)$ which exists only in the interval $[0, \frac{\pi}{2})$. At time $t = \frac{\pi}{2}$ the solution blows up.

The uniqueness property is sometimes violated as well. For instance, the scalar equation

$$\dot{u} = \sqrt{|u|}, \qquad u(0) = 0 \,,$$

has actually infinitely many solutions. Two examples are $u(t) \equiv 0$ and $u(t) = \dfrac{t^2}{4}$. Obviously, the right-hand side does not satisfy the Lipschitz condition which guarantees the uniqueness.

---

[57]Rudolf Otto Sigismund Lipschitz (1832–1903) is a German Mathematician who made contributions to Mathematical Analysis. He studied with Gustav Dirichlet at the University of Berlin.

[58]William Fogg Osgood (1864–1943) was an American Mathematician born un Boston, MA. He studied in Universities of Göttingen and Erlangen, Germany. He made contributions to Mathematical and Complex Analysis, Conformal Mappings.

[59]Charles Émile Picard (1856–1941) was a French Mathematician who made contributions to the Mathematical Analysis. He was married to Marie, a daughter of his Professor Charles Hermite.

Part III
Praxis

Chapter 9
Exercises and Problems

The first two parts of this book provided some theoretical background for solving diffusion problems in building physics and presented traditional (finite differences and finite elements) and nontraditional numerical methods (boundary integral approach, reduced order methods, and spectral methods). In addition, some practical examples were provided to assist the readers in comprehending the methods. This chapter aims at providing more complex exercises and problems to further understand those methods. First, we provide some exercises on the discretization of diffusion equations. Then, heat and mass diffusion problems are presented. Students are also advised to consult the website http://exact.unl.edu/exact/home/home.php for an important library of analytical solutions that may help in validating heat transfer numerical codes developed during their initial training stage.

9.1 Discretization of Diffusion Equations

Consider and recall the 1D diffusion equation:

$$\frac{\partial u}{\partial t} u - \alpha \frac{\partial^2 u}{\partial x^2} = 0 \qquad\qquad x \in \Omega,\, t > 0, \qquad (9.1)$$

where u is the field of interest (temperature, moisture content, or any other diffusion variable) and α is the diffusivity in the domain Ω.

9.1.1 Treatment of the Boundary Conditions

For the domain $\Omega = \left[a,\, b \right]$, discretize the following boundary conditions by using the Backward Time Centered Space (BTCS) scheme (see Sect. 3.1.3.2):

© Springer Nature Switzerland AG 2019
N. Mendes et al., *Numerical Methods for Diffusion Phenomena in Building Physics*,
https://doi.org/10.1007/978-3-030-31574-0_9

9.1.1.1 Neumann-Type

When the Neumann-type boundary conditions are considered,

$$\frac{\partial u}{\partial x}(a, t) = \psi_a,$$

$$\frac{\partial u}{\partial x}(b, t) = \psi_b,$$

1. By considering a *ghost point* x_0 located at the distance $x_0 = a - \Delta x$, show that the Neumann boundary condition at point a can be approximated by

$$\frac{\partial u}{\partial x}(a, t) \simeq -\frac{u_2^{n+1} - u_0^{n+1}}{2\Delta x}.$$

2. By substituting the boundary condition in the discrete heat equation associated to the BTCS discretization, show that the discrete diffusion equation at the boundaries reads as

$$-\text{Fo}\left(u_2^{n+1} + 2\psi_a \Delta x\right) + (1 + 2\text{Fo})\, u_1^{n+1} - \text{Fo}\, u_2^{n+1} = u_1^n \quad x = a,$$

$$-\text{Fo}\, u_{N-1}^{n+1} + (1 + 2\text{Fo})\, u_N^{n+1} - \text{Fo}\, u_{N+1}^{n+1} = u_N^n \quad x = b,$$

with $\text{Fo} = \frac{\alpha \Delta t}{\Delta x^2}$.

3. Show that the matrix system associated to the Neumann boundary condition and the BTCS scheme becomes

$$A\mathbf{u} = f, \tag{9.2}$$

with

$$A = \begin{pmatrix} 1+2\text{Fo} & -2\text{Fo} & 0 & 0 & \cdots & 0 \\ -\text{Fo} & 1+2\text{Fo} & -\text{Fo} & 0 & \cdots & 0 \\ \vdots & & & & & \vdots \\ 0 & \cdots & 0 & -\text{Fo} & 1+2\text{Fo} & -\text{Fo} \\ 0 & \cdots & 0 & 0 & -2\text{Fo} & 1+2\text{Fo} \end{pmatrix} \quad \text{and } f = \begin{pmatrix} u_1^n \\ u_2^n \\ \vdots \\ u_{N-1}^n \\ u_N^n \end{pmatrix} + \begin{pmatrix} 2\text{Fo}\,\psi_a \Delta x \\ 0 \\ \vdots \\ 0 \\ 2\text{Fo}\,\psi_b \Delta x \end{pmatrix}.$$

4. Write the new matrix system, that is, Eq. (9.2), when the fluxes at the boundaries are approximated by a decentered second-order formula given by the following expressions:

$$\frac{\partial u}{\partial x}(x_j, t) \simeq -\frac{-u_{j+2} + 4u_{j+1} - 3u_j}{2\Delta x},$$

$$\frac{\partial u}{\partial x}(x_j, t) \simeq -\frac{-u_{j-2} - 4u_{j-1} + 3u_j}{2\Delta x}.$$

9.1.1.2 Robin-Type

The Robin-type condition corresponds to mixed boundary conditions.

$$\frac{\partial u}{\partial x}(a, t) = \lambda_a u(a, t) + \mu_a ,$$

$$\frac{\partial u}{\partial x}(b, t) = \lambda_b u(b, t) + \mu_b .$$

1. Explicate the matrix system in Eq. (9.2) corresponding to the Robin-type boundary condition and show that the first and last lines become

$$A_{1,1} = 1 + 2\mathsf{Fo} - 2\mathsf{Fo}\Delta x \lambda_a ,$$
$$A_{1,2} = -2\,\mathsf{Fo} ,$$
$$A_{N,N-1} = -2\,\mathsf{Fo} ,$$
$$A_{N,N} = 1 + 2\mathsf{Fo} - 2\mathsf{Fo}\Delta x \lambda_b .$$

and the RHS of the system in Eq. (9.2) becomes

$$f_1 = u_1^n + 2\mathsf{Fo}\Delta x \mu_a ,$$
$$f_N = u_N^n + 2\mathsf{Fo}\Delta x \mu_b .$$

9.1.2 Numerical Solution

2. Write down the tridiagonal matrix system corresponding to the Crank–Nicolson scheme for 1D heat transfer using the Robin boundary conditions.
3. Compute the solution of Eq. (9.1) by considering the following numerical application:

$$a = 0 \qquad b = 1 \qquad \lambda_a = 0.5 \qquad \lambda_b = 0.9$$
$$\mu_a = 1 \qquad \mu_b = -5 \qquad u(x, 0) = u_{in} = 1$$

The solution can be compared with the analytical solution provided by the EXACT group[1] X33B11T1G1. The Matlab code is recalled here, adapted using the current notation.

---

[1] http://exact.unl.edu/exact/home/home.php. In general, students are advised to consult this website and its content offering an important library of analytical solutions that may help validate the developed numerical codes.

```
1    function X33B11T1G1(lambda_a, lambda_b, mu_a, mu_b, u_in)
2
3    g    = 0;
4    bi1  = lambda_a;
5    bi2  = lambda_b;
6    q1   = mu_b;
7    q2   = mu_a;
8    Tf1  = 0;
9    Tf2  = 0;
10   Ti   = u_in;
11   tm0  = 0.01;
12
13
14   nterm = 1+sqrt(round(23/tm0))/pi;
15   b12    = bi1+bi2;
16   if (b12>0)
17   n1 = 1;
18   else
19   n1 = 2;
20   beta(1) = 0;
21   end
22   for n = n1:nterm
23   cn   = (n-3/4)*pi;
24   dn   = (n-1/2)*pi;
25   d    =(122*(n-1)+79)/100;
26   p23  = 1/n;
27   p13  = 1-9/10/n;
28   bx   = bi1*bi2;
29   bs   = bi1+bi2;
30   gam  = 1-(104/100)*(sqrt((bs+bx+cn-pi/4)/(bs+bx+d))-...
31   (bs+bx+sqrt(d*(cn-pi/4)))/(bs+bx+d));
32   x23  = (bi1+bi2-cn)/(bi1+bi2+cn);
33   x13  = bx/(bx+2/10+bs*(pi*pi*(n-5/10)/2));
34   g13  = 1+x13*(1-x13)*(1-85/100/n-(6/10-71/100/n)*(x13+1)*(x13-6/10-1/4/n));
35   e23  = gam*(p23*x23+(1-p23)*tanh(x23)/tanh(1));
36   e13  = 2*g13*x13;
37   y23  = cn+pi*e23/4;
38   y13  = dn+pi*e13/4;
39   zn   = (y13*bx/(bx+dn^2)+y23*(1-bx/(bx+dn^2)));
40   a    = (1+bs)*sin(zn)+2*zn*cos(zn);
41   b    = (zn^2-bs-bx)*cos(zn)+(2+bs)*zn*sin(zn);
42   c    = (zn^2-bx)*sin(zn)-bs*zn*cos(zn);
43   e    = -c/b-(-cos(zn)*c^3+a*c^2*b)/(3*cos(zn)*c*c*b-2*a*c*b^2+b^4);
44   z1   = zn+e;
45   a1   = (1+bs)*sin(z1)+2*z1*cos(z1);
46   b1   = (z1^2-bs-bx)*cos(z1)+(2+bs)*z1*sin(z1);
47   c1   = (z1^2-bx)*sin(z1)-bs*z1*cos(z1);
48   e1   = ...
                -c1/b1-(-cos(z1)*c1^3+a1*c1^2*b1)/(3*cos(z1)*c1*c1*b1-2*a1*c1*b1^2+b1^4);
49   z2   = z1+e1;
50   a2   = (1+bs)*sin(z2)+2*z2*cos(z2);
51   b2   = (z2^2-bs-bx)*cos(z2)+(2+bs)*z2*sin(z2);
52   c2   = (z2^2-bx)*sin(z2)-bs*z2*cos(z2);
53   e2   = ...
                -c2/b2-(-cos(z2)*c2^3+a2*c2^2*b2)/(3*cos(z2)*c2*c2*b2-2*a2*c2*b2^2+b2^4);
54   z3   = z2+e2; beta(n)=z3;
55   end
56   if n1 == 1
57   gs   = g;
58   else
59   gs   = q2-q1;
60   end
61   Co1  = bi1*Tf1+q1; Co2=bi2*Tf2-q2;
62   if n1==1
63   A    = (-2*bi2*Co1+2*bi1*Co2+bi1*(2+bi2)*g)/(2*(bi1+bi2+bi1*bi2));
64   else
65   A    = -q1;
66   end
67   if n1==1
68   B    = (2*(1+bi2)*Co1+2*Co2+(2+bi2)*g)/(2*(bi1+bi2+bi1 *bi2));
```

```
69   else
70   B     = (2*q1+q2)/6;
71   end
72   for n=n1:nterm
73   Nrm(n) = ((beta(n)^2+bi1^2)*(1+bi2/(beta(n)^2+bi2^2))+bi1)/2;
74   Pn(n)  = bi1*gs*(1-cos(beta(n)))/beta(n)^3+(A*bi1+gs-gs*bi1)...
75          *sin(beta(n))/beta(n)^2+(-A+bi1*(B-Ti)+(A-A*bi1-B*bi1-gs+...
76          gs*bi1/2+bi1*Ti)*cos(beta(n)))/beta(n)+(A+B-gs/2-Ti)*sin(beta(n)));
77   end
78
79   % post process
80   xres      = linspace(0,1,50);
81   PlotTimes = [0.05,0.1,0.2,0.5,1.0];
82   Colors    = ['k','b','r','g','m','c','y'];
83   Nx        = length(xres);
84   xi        = zeros(length(PlotTimes),Nx);
85   yi        = xi;
86   for i=1:length(PlotTimes)
87   xi(i,:)   = xres;
88   yi(i,:)   = tempfunc(xres,PlotTimes(i),beta,Pn,Nrm,bi1,n1,A,B,g,gs,q1,q2,Ti);
89   end
90   figure; hold on;
91   MyLegend = {};
92   for i=1:length(PlotTimes)
93   plot(xi(i,:),yi(i,:),Colors(i),'linewidth',2);
94   MyLegend{i} = num2str(PlotTimes(i));
95   end
96   plot(xres,steadytemp(xres,max(PlotTimes)+0,n1,A,B,g,gs,q1,q2,Ti),...
97   '--k','linewidth',2);
98   xlabel('x')
99   ylabel('u')
100  title('Field')
101  legend(horzcat(MyLegend,'steady state'),'Location','Best');
102  hold off;
103  end
```

The function X33B11T1G1 uses the following subfunctions:

```
1    function [Tmp] = ...
         tempfunc(x,tm,beta,Pn,Nrm,bi1,n1,A,B,g,gs,q1,q2,Ti)
2    x    = x(:);
3    if n1==1
4    Tmp = 0;
5    else
6    Tmp = A/2+B-gs/6;
7    end
8    Xn   = zeros(length(x),length(beta));
9    for n=n1:length(beta)
10   Xn(:,n) = beta(n).*cos(beta(n).*x)+bi1*sin(beta(n).*x);
11   Tmp = Tmp+Pn(n)*Xn(:,n)*exp(-beta(n)^2*tm)/Nrm(n);
12   end
13   Tss = steadytemp(x,tm,n1,A,B,g,gs,q1,q2,Ti);
14   Tmp = Tss-Tmp;
15   end
```

```
1    function [Tss] = steadytemp(x,tm,n1,A,B,g,gs,q1,q2,Ti)
2        Tss = (n1-1)*(Ti+g*tm+(q1-q2)*tm)-gs*x.^2/2+A.*x+B;
3    end
```

9.2   Heat and Mass Diffusion: Numerical Solution

This exercise focuses on the numerical solution of the coupled equation. Let us consider the system based on [2]:

$$\rho C_p \frac{\partial T}{\partial t}(x,t) = \kappa \frac{\partial^2 T}{\partial x^2}(x,t) + \rho C_m (\varepsilon h_{lv} + \gamma) \frac{\partial m}{\partial t}(x,t), \qquad (9.3a)$$

$$\rho C_m \frac{\partial m}{\partial t}(x,t) = D_m \frac{\partial^2 m}{\partial x^2}(x,t) + \delta D_m \frac{\partial^2 T}{\partial x^2}(x,t), \qquad (9.3b)$$

where $T(x,t)$ and $m(x,t)$ are heat and mass content fields respectively, x is the space coordinate between 0 and 1, and t represents the time variable. The physical parameters ρ, C_p, C_m, κ, D_m, ε, h_{lv}, δ, and γ must be constant. The boundary conditions are as follows:

for $x = 0$: $\kappa \frac{\partial T}{\partial x}(0,t) = h\left(T(0,t) - T^\star\right), \quad D_m \frac{\partial m}{\partial x}(0,t) = h_m\left(m(0,t) - m^\star\right),$

$$\hspace{11cm} (9.4a)$$

for $x = 1$: $\frac{\partial T}{\partial x}(1,t) = 0, \hspace{3cm} \frac{\partial m}{\partial x}(1,t) = 0, \hspace{1.5cm} (9.4b)$

where $T^\star$ and $m^\star$ are fixed and known quantities.

1. Let us recast these equations to more conveniently apply the finite difference approach. Find a suitable change of variables for Eq. (9.3a) to be recast in

$$\frac{\partial T}{\partial t}(x,t) - v \frac{\partial m}{\partial t}(x,t) = L \frac{\partial^2 T}{\partial x^2}(x,t), \qquad (9.5a)$$

$$\frac{\partial m}{\partial t}(x,t) - \lambda \frac{\partial T}{\partial t}(x,t) = D \frac{\partial^2 m}{\partial x^2}(x,t). \qquad (9.5b)$$

2. Show that the system of equations (9.5a) has the following matrix formulation, when $\lambda v \neq 1$:

$$\frac{\partial}{\partial t}\begin{pmatrix} T(x,t) \\ m(x,t) \end{pmatrix} = \frac{1}{1-\lambda v}\begin{bmatrix} L & vD \\ \lambda L & D \end{bmatrix}\frac{\partial^2}{\partial x^2}\begin{pmatrix} T(x,t) \\ m(x,t) \end{pmatrix} \qquad (9.6a)$$

given that

$$\begin{bmatrix} 1 & -v \\ -\lambda & 1 \end{bmatrix}^{-1} = \frac{1}{1-\lambda v}\begin{bmatrix} 1 & v \\ \lambda & 1 \end{bmatrix}.$$

3. Now, we construct a fully implicit scheme for Eq. (9.6), by denoting

$$\mathbf{u}(x,t) = \begin{pmatrix} T(x,t) \\ m(x,t) \end{pmatrix} \hspace{2cm} \mathscr{A} = \frac{1}{1-\lambda v}\begin{bmatrix} L & vD \\ \lambda L & D \end{bmatrix}$$

such that Eq. (9.6) becomes

$$\frac{\partial \mathbf{u}}{\partial t}(x, t) = \mathscr{A}\frac{\partial^2 \mathbf{u}}{\partial x^2}(x, t). \tag{9.7}$$

Set a spatial discretization of $[0, 1]$ by using $N + 1$ points regularly spaced by dx such that the first point x_0 is located at $x = 0$ and the last point x_N is at $x = 1$. Note that $\mathbf{u}_j^n$, T_j^n, and m_j^n are the approximations of $\mathbf{u}$, T, and m, respectively, and are evaluated at points jdx, for $j = 0, \ldots, N$, at time ndt, for $n \geq 0$.

a. Show that a fully implicit scheme, with a central difference approximation for the diffusion terms, reads as

$$-\alpha\mathscr{A}\,\mathbf{u}_{j+1}^{n+1} + (1 + 2\alpha\mathscr{A})\,\mathbf{u}_j^{n+1} - \alpha\mathscr{A}\,\mathbf{u}_{j-1}^{n+1} = \mathbf{u}_j^n \tag{9.8}$$

with

$$\alpha = \frac{vdt}{dx^2}.$$

b. Recall the Robin-type boundary conditions in Eq. (9.4a):

$$\kappa\frac{\partial T}{\partial x}(0, t) = h\left(T(0, t) - T^\star\right)$$

$$D_m\frac{\partial m}{\partial x}(0, t) = h_m\left(m(0, t) - m^\star\right).$$

Consider a second-order approximation for these equations involving a ghost point for T_{-1} and m_{-1} and show what the numerical scheme for $\mathbf{u}$ yields for the first node $j = 0$ in

$$\mathbf{u}_{-1}^{n+1} = \mathbf{u}_1^{n+1} - 2dx\,\Gamma\left(\mathbf{u}_0^{n+1} - \mathbf{u}^\star\right),$$

with

$$\Gamma = \begin{bmatrix} \frac{h}{\kappa} & 0 \\ 0 & \frac{h_m}{D_m} \end{bmatrix}.$$

c. By inserting the Robin-type discrete equation into the heat equation, show that Eq. 9.8 at the first node becomes

$$-2\alpha\mathscr{A}\,\mathbf{u}_1^{n+1} + (1 + 2\alpha\mathscr{A}\,(1 + dx\,\Gamma))\,\mathbf{u}_0^{n+1} = \mathbf{u}_0^n + \alpha\mathscr{A}\,2dx\,\Gamma\mathbf{u}^\star.$$

d. For the last node $j = N$, show the result obtained by the boundary condition in Eq. (9.4b) in $\mathbf{u}_{N+1}^{n+1} = \mathbf{u}_{N-1}^{n+1}$.

e. By using a ghost node $\mathbf{u}_{N+1}^{n+1}$, and substituting it into the implicit scheme expressed at $j = N$, show that Eq. (9.8) reads as:

$$(1 + 2\alpha \mathscr{A}) \, \mathbf{u}_N^{n+1} - 2\alpha \mathscr{A} \, \mathbf{u}_{N-1}^{n+1} = \mathbf{u}_N^n.$$

f. Show that the final matrix system can be formulated as follows:

$$\mathscr{M} \mathbf{u}^{n+1} = f,$$

where $\mathscr{M}$ is a $(2(N + 1)) \times (2(N + 1))$ tridiagonal-block matrix, in which each block-matrix $M_{i,j}$ is of dimension 2×2.

$$\mathscr{M} = \begin{bmatrix} M_{00} & M_{01} & 0 & \cdots & & & & 0 \\ M_{10} & M_{11} & M_{12} & 0 & \cdots & & & \\ 0 & \ddots & \ddots & \ddots & & & \cdots & \vdots \\ \vdots & 0 & M_{j,j-1} & M_{j,j} & M_{j,j+1} & & 0 & \\ & & & \ddots & \ddots & & \ddots & 0 \\ \vdots & & \cdots & 0 & M_{N-1,N-2} & M_{N-1,N-1} & M_{N-1,N} \\ 0 & & & \cdots & & 0 & M_{N,N-1} & M_{N,N} \end{bmatrix}.$$

The vectors $\mathbf{u}$ and f are of the dimension $2(N + 1)$ as:

$$\mathbf{u}^{n+1} = \begin{pmatrix} u_0^{n+1} \\ \vdots \\ u_{j-1}^{n+1} \\ u_j^{n+1} \\ u_{j+1}^{n+1} \\ \vdots \\ u_N^{n+1} \end{pmatrix} = \begin{pmatrix} T_0^{n+1} \\ m_0^{n+1} \\ -- \\ \vdots \\ T_{j-1}^{n+1} \\ m_{j-1}^{n+1} \\ -- \\ T_j^{n+1} \\ m_j^{n+1} \\ -- \\ T_{j+1}^{n+1} \\ m_{j+1}^{n+1} \\ -- \\ \vdots \\ T_N^{n+1} \\ m_N^{n+1} \end{pmatrix} \quad \text{and} \quad f = \begin{pmatrix} u_0^n \\ \vdots \\ u_{j-1}^n \\ u_j^n \\ u_{j+1}^n \\ \vdots \\ u_N^n \end{pmatrix} + \begin{pmatrix} \Psi_0(\Gamma, \bar{u}^\star) \\ \vdots \\ 0 \\ 0 \\ 0 \\ \vdots \\ 0 \end{pmatrix}.$$

4. Write a computer code to solve the aforementioned problem using the following data in S.I.:

$$C_p = 2500 \quad \kappa = 0.65 \quad \rho = 370 \quad C_m = 0.01 \quad D_m = 2.2 * 10^{-8} \quad h_{lv} = 2500 \quad \varepsilon = 0.3 \quad \delta = 2.$$

The initial and boundary conditions are:

$$T^* = 60\,°C \quad m^* = 45\,°M \quad T(t = 0) = 10\,°C \quad m(t = 0) = 86\,°M$$

The length of the material is $l = 0.024\,m$ and the solution can be computed at $t = 150\,s$.

5. Reconsider the exercise using a central-difference scheme and an explicit method. Perform a parametric analysis on the grid refinement and time step effects.
6. Reconsider the exercise using a MultiTridiagonal Matrix Algorithm (MTDMA) (See Sect. (2.8)) and compare the advantages and disadvantages between the two algorithms.
7. Reconsider the exercise considering 3D transfer.

Remark

The previously solved problem can be considered for other advanced numerical methods, particularly the model reduction and spectral methods. The following analytical solution can be used to verify the computer code. It is obtained from [2]. Note that it has been written for Dirichlet boundary conditions.

```
1   %-----------------------------------------------------
2   % Analytical solution
3   % Contribution to analytical and numerical study of combined heat
4   % and moisture transfers in porous building materials
5   % K. Abahri, R. Belarbi, A. Trabelsi (2010)
6   %-----------------------------------------------------
7
8   clear all
9   close all
10
11  % Material properties
12  C    = 2500;              % specific heat
13  k    = 0.65;              % thermal conductivity
14  ro   = 370;               % material density
15  Cm   = 0.01;              % moisture storage capacity
16  Dm   = 2.2*10^(-8);       % Moisture diffusion
17  hlv  = 2500;              % heat of phase change
18  epsilon = 0.3;            % ratio of vapor diffusion coefficient to total ...
                moisture diffusion coefficient
19  Δ    = 2;                 % thermogradient coefficient
20
21  % Initial and boundary conditions
22  Ti = 10;                  % initial temperature
23  T1 = 60;                  %left side temperature
24  T2 = 110;                 %right side temperatue
25  mi = 86;                  %initial humidity
26  m1 = 45;                  %left side humidity
27  m2 = 4;                   %right side humidity
28
29  % Space domain
30  l   = 0.024;              % length material
31  dx  = 0.0002;             % space mesh
32  nx  = l/dx;               % mesh numbers
```

```
33   x  = linspace(0,1,nx);  % spatial grid
34
35   % simulation time
36   t = 150;                 % time
37
38   % Simplifications
39   L = k/(ro*C);
40   D = (k*Dm)/((ro*Cm)*(k+Dm*Δ*epsilon*hlv));
41   v = Cm*epsilon*hlv/C;
42   lambda = (C*Dm*Δ)/(Cm*(k+Dm*Δ*epsilon*hlv));
43
44   % Coefficients
45   q1=((-1)/sqrt(2*D))*sqrt(1+(D/L)-sqrt((1-D/L)^2)+...
46       (4*lambda*v*D)/L);
47   q2=((-1)/sqrt(2*D))*sqrt(1+(D/L)+sqrt((1-D/L)^2)+...
48       (4*lambda*v*D)/L);
49
50   A1=((m2-mi)*v-(1-L*q2*q2)*(T2-Ti))/(v*(q2*q2-q1*q1));
51   A2=(-(m2-mi)*v+(1-L*q1*q1)*(T2-Ti))/(v*(q2*q2-q1*q1));
52   B1=(-(m1-mi)*v+(1-L*q2*q2)*(T1-Ti))/(v*(q2*q2-q1*q1));
53   B2=((m1-mi)*v-(1-L*q1*q1)*(T1-Ti))/(v*(q2*q2-q1*q1));
54
55   % Series
56   sa1=0; sa2=0; sb1=0; sb2=0; %series
57
58   % Check the convergence of the finite sum
59   for n=1:30
60     sa1 = sa1+(((-1)^n/n)*exp((-n*n*pi*pi*t)/(q1*q1*l*l))*sin(n*pi*x/l));
61     sa2 = sa2+(((-1)^n/n)*exp((-n*n*pi*pi*t)/(q2*q2*l*l))*sin(n*pi*x/l));
62     sb1 = sb1+(((-1)^n/n)*exp((-n*n*pi*pi*t)/(q1*q1*l*l))*sin(n*pi*(x-1)/l));
63     sb2 = sb2+(((-1)^n/n)*exp((-n*n*pi*pi*t)/(q2*q2*l*l))*sin(n*pi*(x-1)/l));
64   end
65
66   % Analytical solution of temperature and moisture
67
68   T = v/L*((A1*x/l+2*A1/pi*sa1+B1*(x-1)/l+2*B1/pi*sb1)+...
69       (A2*x/l+2*A2/pi*sa2+B2*(x-1)/l+2*B2/pi*sb2))+...
70       Ti;
71   m = ((1/L-q1*q1)*(A1*x/l+2*A1/pi*sa1+B1*(x-1)/l+2*B1/pi*sb1))+...
72       ((1/L-q2*q2)*(A2*x/l+2*A2/pi*sa2+B2*(x-1)/l+2*B2/pi*sb2))+...
73       mi;
74
75   % Postprocess
76   figure
77   subplot(1,2,1)
78   plot(x,T,'rx-')
79   xlabel('x')
80   ylabel('temperature')
81
82   subplot(1,2,2)
83   plot(x,m,'bx-')
84   xlabel('x')
85   ylabel('moisture')
```

9.3   Whole Building Energy Simulation

9.3.1   *Heat Transfer*

This exercise involves the coupling between the diffusion transfer in the porous walls and the heat and moisture balances in the room air as illustrated in Fig. 9.1. This case study was inspired by the benchmark BESTEST [12, 200]. It is assumed that the

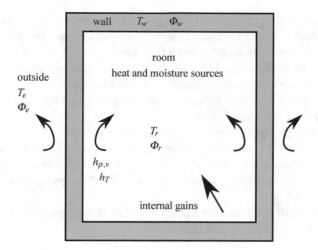

Fig. 9.1 Illustration of the case study

Table 9.1 Dry-basis properties of wall materials

Soil	ρ_0 (m³/kg)	c_m (J/kg.K)	λ_m (W/m.K)
Aerated concrete	650	840	0.18
EPS	10	1450	0.05
Concrete	2500	1000	1

floor, roof, and walls are composed of aerated concrete, the properties of which are given in Table 9.1. The area of the walls is 75.6 m², and that for the roof and floor is 48 m² each. Their thicknesses are 0.15 cm. Radiative transfer is not considered, and the inside and outside surface coefficients are

$$h_{T,i} = 8.29 \qquad \qquad (\text{W}/\text{m}^2.\text{K}),$$
$$h_{T,e} = 29.30 \qquad \qquad (\text{W}/\text{m}^2.\text{K}).$$

A 100% convective heat gain of 500 W between 9 am and 5 pm and a 24 h ventilation rate of 0.5 ach are considered. The results should be presented for the 30th day of simulation to prevent effects imposed by the assumed initial values of temperature (20 °C) and a sinus function to represent the daily variation of external temperature:

$$T_e = 298.15 + \sin(2\pi t/86400) \qquad \qquad (\text{K})$$

1. Define the governing differential equations of the wall heat diffusion problem and of the inside of room as balanced. Compute the temperature inside the room.
2. Verify the effects of time step and grid refinement for heat transfer.

3. Add an Expanded polystyrene (EPS) insulation material of 5 cm (properties are given in Table 9.1) on the external side of the building envelope, and compare the temperature profile with and without the material.
4. Repeat exercise (3) by placing the EPS layer at the internal boundary.
5. Repeat exercise (3) by considering the EPS on both sides.
6. Now, consider a 5 cm EPS at the external side and a 3 cm concrete (high thermal mass) layer on the internal side.
7. Repeat exercise (6) by inverting the order of layers.
8. Elaborate a table to summarize the results of Exercises (2)–(7) in terms of daily energy transferred, internal temperature peaks, and delay and damping effects. Compare your results presented with those obtained by running a building-simulation software.
9. What could be the minimum warm-up period so that the initial conditions do not affect the tested different materials?

9.3.2   Moisture Transfer

Consider the same building model as described in the previous exercise but isother-mally at 20 °C. A 500 g/h indoor moisture production between 9 am and 5 pm, and a 24 h ventilation rate of 0.5 ach is considered. The moisture properties of the aerated concrete, vapor permeability, and sorption isotherm are as follows:

$$\delta_v = 3 \cdot 10^{-11} \qquad\qquad (s)$$
$$\theta = 0.042965\phi$$

The mass convective coefficient based on vapor pressure difference is $h_m = 0.00275$ m/s for the inside and outside parts of the wall. The outside conditions for relative humidity are constant and at 30%. The results should be presented for the 365th day of simulation to prevent effects imposed by the assumed initial values of moisture content as moisture diffuses slowly.

1. Define the governing differential equations of the wall and the inside of room. In addition, compute the relative humidity inside the room.
2. Conduct a sensitivity analysis on the effects of time step and grid refinement for moisture transfer.
3. Compare the daily relative humidity profile with and without adsorption.
4. Compare your results with the results from whole-building simulation programs that consider moisture.
5. What could be the minimum warm-up period to avoid the effects of the initial conditions?

9.3.3 Heat and Moisture Transfer

Now, both the previous exercises can be considered for coupled heat and moisture transport phenomena, including the latent heat effect. Solve the combined problem by using a MTDMA for a central-difference scheme and a fully implicit method.

1. Write the governing equations for both moisture content and vapor pressure driving potentials.
2. Write a computer code for both potentials.
3. Quantify the importance of considering moisture by comparing the result with the results of the previous exercises.
4. How can you consider the effect of rain?
5. Include the room-air governing equation(s) as part of a directly coupled solution by using TDMA/MTDMA, and describe the advantages of this solution to solve the Sects. 9.3.1, 9.3.2 and 9.3.3.
6. Show quantitatively the effects of a latex paint (see values in Chap. 2) on the results of Exercise 3.
7. In your opinion, what are the main challenges in building physics regarding the use of traditional numerical methods applied to diffusion phenomena?

9.4 Heat and Mass Diffusion: Analysis of the Physical Behavior

This exercise deals with the analysis of the physical behavior of the 3D heat and mass diffusion transfer in soils (see Sect. 2.2):

$$\rho c_m(T, \theta) \frac{\partial T}{\partial t} = \nabla \cdot (\lambda(T, \theta) \nabla T) - L(T) \nabla \cdot \mathbf{j}_v ,$$

$$\frac{\partial \theta}{\partial t} = -\nabla \cdot \left(\frac{\mathbf{j}}{\rho_l} \right) ,$$

with

$$\frac{\mathbf{j}}{\rho_l} = - \left(D_T(T, \theta) \frac{\partial T}{\partial x} - D_\theta(T, \theta) \frac{\partial \theta}{\partial x} \right) \mathbf{e_i} - \left(D_T(T, \theta) \frac{\partial T}{\partial y} - D_\theta(T, \theta) \frac{\partial \theta}{\partial y} \right) \mathbf{e_j} ,$$
$$- \left(D_T(T, \theta) \frac{\partial T}{\partial z} - D_\theta(T, \theta) \frac{\partial \theta}{\partial z} + K_l \right) \mathbf{e_k} .$$

Figure 9.2 illustrates the case study, in which the total simulation time is 1 year. Tables 9.2 and 9.3 list the hygrothermal properties of the domain. In terms of boundary conditions, all faces of the cube are considered adiabatic and impermeable (Neumann boundary conditions), except at $z = 0$, where

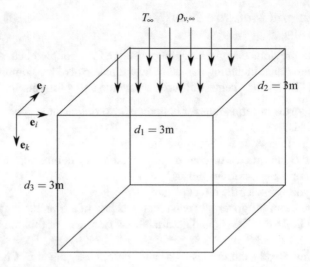

Fig. 9.2 Illustration of the case study

Table 9.2 Dry-basis properties of soils

Soil	ρ_0 (m^3/kg)	c_m (J/kg/K)	λ_m (W/m/K)
Sandy silt	1280	880	0.3452

$$
-\left[\lambda\left(T,\theta\right)\frac{\partial T}{\partial x}-L\,\mathbf{j}_v\right]_{z=0}=
$$
$$
h_T\left(T_\infty-T_{x=0}\right)+Lh_{\rho v}\left(\rho_{v,\infty}\left(T,\theta\right)-\rho_{v,z=0}\left(T,\theta\right)\right),
$$
$$(9.10)$$

$$
-\left[\frac{\partial}{\partial x}\left(\left(\mathrm{D}_{\theta_l}\left(T,\theta\right)+\mathrm{D}_{\theta_v}\left(T,\theta\right)\right)\frac{\partial\theta}{\partial x}+\left(\mathrm{D}_{T_l}\left(T,\theta\right)+\mathrm{D}_{T_v}\left(T,\theta\right)\right)\frac{\partial T}{\partial x}\right)\right]_{z=0}
$$
$$
=\frac{h_{\rho v}}{\rho_l}\left(\rho_{v,\infty}\left(T,\theta\right)-\rho_{v,z=0}\left(T,\theta\right)\right).
$$
$$(9.11)$$

with

$$
T_\infty=288.15+5\sin(2\pi t/31536000)+5\sin(2\pi t/86400)\qquad\text{(K)}
$$
$$
\phi_\infty=0.7+0.1\sin(2\pi t/86400)\qquad\text{(--)}
$$
$$
h_T=5\qquad\text{(W/m}^2\text{/K)}
$$
$$
h_{\rho v}=5\cdot10^{-3}\qquad\text{(m/s)}
$$

1. Write a computer code to solve the problem using the finite-difference approach (or advanced methods) and a fully implicit method.

Table 9.3 Variation of the properties of the Sandy Silt soil with the moisture content [147]

θ (m^3/m^3)	ϕ $(adim.)$	λ (W/m/K)	$D_{\theta\ell}$ (m^2/s)	$D_{T\ell}$ $(m^2/s/K)$	$D_{\theta v}$ (m^2/s)	D_{Tv} $(m^2/s/K)$
0	4.00E-03	3.00E-01	0	0	1.63E-10	9.50E-14
2.59E-02	1.96E-01	3.10E-01	0	0	2.19E-09	3.71E-12
5.17E-02	6.10E-01	3.30E-01	1.10E-31	4.70E-36	2.10E-09	1.16E-11
7.76E-02	8.60E-01	3.50E-01	3.26E-11	1.38E-15	9.04E-10	1.70E-11
1.034E-01	8.68E-01	3.90E-01	2.43E-09	9.26E-14	3.34E-10	1.97E-11
1.293E-01	8.76E-01	4.60E-01	1.30E-08	7.46E-13	7.70E-11	2.13E-11
1.551E-01	8.85E-01	6.00E-01	4.32E-08	3.28E-12	2.65E-11	2.23E-11
1.810E-01	8.93E-01	8.10E-01	1.09E-07	1.03E-11	1.14E-11	2.33E-11
2.068E-01	9.01E-01	1.11E+00	2.35E-07	2.64E-11	5.67E-12	2.42E-11
2.327E-01	9.09E-01	1.40E+00	4.52E-07	5.83E-11	3.09E-12	2.52E-11
2.585E-01	9.18E-01	1.59E+00	8.03E-07	1.16E-10	1.80E-12	2.61E-11
2.844E-01	9.26E-01	1.66E+00	1.34E-06	2.12E-10	1.10E-12	2.70E-11
3.102E-01	9.34E-01	1.67E+00	2.17E-06	3.65E-10	6.98E-13	2.78E-11
3.361E-01	9.42E-01	1.67E+00	3.41E-06	5.96E-10	4.53E-13	2.87E-11
3.619E-01	9.51E-01	1.68E+00	5.27E-06	9.29E-10	2.98E-13	2.95E-11
3.878E-01	9.59E-01	1.68E+00	8.12E-06	1.39E-09	1.97E-13	3.03E-11
4.136E-01	9.67E-01	1.68E+00	1.26E-05	2.02E-09	1.29E-13	3.10E-11
4.395E-01	9.75E-01	1.68E+00	2.04E-05	2.83E-09	8.21E-14	3.15E-11
4.653E-01	9.84E-01	1.68E+00	3.54E-05	3.77E-09	4.87E-14	3.15E-11
4.912E-01	9.92E-01	1.68E+00	7.47E-05	4.55E-09	2.38E-14	3.02E-11
5.169E-01	1.00E+00	1.68E+00	1.93E-03	2.69E-10	2.14E-16	4.38E-13

2. Reconsider the exercise by using the model described in Chap. 2, Sect. 2.2, and Eqs. (2.14) and (2.15). The model can be analyzed and compared for high relative humidity and/or by considering a source term $\mathbf{j}_{rain}$ at the upper surface (Eq. (9.11)):

$$\mathbf{j}_{rain} = 5 \cdot 10^{-4} \qquad (kg/m^2.s) \qquad \forall t \in [7, 12]h$$
$$\mathbf{j}_{rain} = 0 \qquad (kg/m^2.s) \qquad \forall t \notin [7, 12]h$$

Chapter 10
Conclusions

In the first part of this book, we presented a practical introduction on diffusion phenomena in building physics and traditional numerical methods for solving their governing partial differential equations.

Heat diffusion models and numerical methods have been extensively used and improved in building simulation tools since the early seventies; however, it was only in the nineties that mass diffusion began to be included for predicting accurate conduction loads and assessing risks on mold growth, interstitial condensation, material deterioration, and indoor air quality degradation. However, after more than 30 years, the diffusion models, through building envelopes, can still be considered as simplified and, under uncertain conditions, not sufficiently accurate for the analysis of the hygrothermal performance of whole buildings.

The diffusion phenomena in building elements are not simple. They are three-dimensional and the fields of temperature- and mass-related variables, such as air and vapor pressures, relative humidity, moisture content, and capillary suction pressure, are directly dependent on the air flow and radiation asymmetric patterns from indoors and outdoors, as well as the complexity imposed by driving rain and snow as boundary conditions. Furthermore, many other difficulties can be added, including (*i*) anisotropy of some porous materials, (*ii*) advective terms for heat and mass transfer; (*iii*) scattered radiative heat transfer through insulating materials; (*iv*) contact resistance between material layers; (*v*) dimensional changes due to shrinking and swelling; (*vi*) time-dependent hygrothermal properties; (*vii*) ice formation; and (*viii*) crack formation.

Moreover, hygrothermal models require transport coefficents and properties strongly dependent on the moisture content, and their behavior at high relative humidity remains very uncertain. In addition, very small changes in relative humidity or vapor pressure may extensively change the values of transport coefficients, considerably complicating the numerical solution of a combined heat and mass diffusion problem, thus imposing risks of increasing inaccuracy or even causing numerical divergence.

© Springer Nature Switzerland AG 2019   229
N. Mendes et al., *Numerical Methods for Diffusion Phenomena in Building Physics*,
https://doi.org/10.1007/978-3-030-31574-0_10

Some strategies associated to heat and mass diffusion models are presented in Chap. 2; however, next generations of PhD students may have to face many great challenges regarding diffusion phenomena in building physics.

Traditional numerical methods largely used in building simulation tools are presented in the first part of this book, in a practical manner. Chapter 3 presents the Finite-Difference Method (FDM) largely used for solving either heat or Heat–Air–Moisture (HAM) transfer problems through building elements. The FDM is simple to use but limited to more simplified geometries.

Although the Finite-Element Method (FEM), presented in Chap. 4, is extensively used by the heat–and–fluid mechanics communities, its adoption in building physics is still humble. Important characteristics, such as robustness, flexibility, and adaptability to any type of geometry are encouraging factors in the applications of the FEM in building physics computer codes. FEM has evolved in the last decades and can solve advective problems, allowing for its integrated use for solving room air flow equations.

Nowadays, the FEM can be compared with the finite-volume method (not presented in this book) for solving multiphysics problems. Theoretically, the numerical analysis is also well established because the numerical properties are well understood, thus providing a high level of confidence. Therefore, compared with the FDM, the FEM proposes more powerful aspects; however, much more time is required for understanding abstract concepts and the required computational technical skills, at least for multidimensional cases.

Currently, in most building physics tools, diffusion phenomena are modeled in one space dimension. However, recent developments, such as asymmetric short- and long-wave radiation calculated by advanced computer graphics techniques, that consider more realistic effects associated, for instance, with radiative boundary conditions expect the diffusion modeling to go beyond the assumptions currently used. The inclusion of generic geometries and spatial heterogeneities, for example, reveals that these are parameters of prime importance that are better supported by the FEM than the FDM.

As multidimensional combined heat and moisture transfer in large domains of building physics, or even in urban physics, is a trend for the near future, research on advanced numerical methods is necessary. Thus, in the second part of this book, we present numerical methods that we believe are promising in building physics by avoiding those limitations especially concerned with the simulation cost.

The first group of advanced methods is described in Chap. 5 where some strategies to overcome the CFL stability restrictions are proposed in the framework of explicit (in time) schemes. The second advanced method introduced is associated with model reduction techniques. The balanced truncation, the modal identification, the Proper Orthogonal Decomposition (POD) and the Proper Generalised Decomposition (PGD) were all presented in Chap. 7. The Reduced Order Model (ROM) methods enable the determination of the solution of a problem in a reduced space dimension and are even more promising for 2D or 3D diffusion models that are increasingly appearing in this literature field. Among the advanced methods, PGD is so far the most promising, with more available references applied to building physics.

Some practical exercises are provided to help the readers to apply the methods as well as to illustrate the benefits of using them. Research is encouraged to expand the advanced numerical methods for solving HAM coupled equations presented in Chap. 2.

In some building physics problems, the relevant information lies at the surface of a domain. In addition, only a few point-wise evaluations may be needed; this motivated the presentation of boundary integral approaches, also known as meshless techniques, in Chap. 7. They present conceptually greater intrinsic advantages: the number of degrees of freedom is drastically reduced; thus, the computational cost may be theoretically suitable for simulation over large spatial 3D domains (e.g., for diffusion transfer into the soil). However, these approaches seem to have severe limitations when applied to real problems in buildings: the implementation of Robin-type boundary conditions or the inclusion of a source term induces specific numerical treatments that can significantly degrade the gain in time. More fundamentally, Green functions and Trefftz complete functions are available only for a few types of linear equations. The consequence of this limitation complicates the application of such techniques to realistic configurations for which the physical properties of the medium are not constant. However, meshless approaches might be used in the future, as their use can be combined with other methods, such as domain decomposition and multigrid methods, to present perspectives of great interest for building physics applications. The main concept of the boundary integral approach is to reduce the domain of calculation by using an analytic expression of the unknown field, involving surface computation.

Chapter 8 presented another advanced numerical method. The chapter focused on Spectral Methods (SM), presented some theoretical bases for spectral discretization, and illustrated an application to a problem stemming from this field. Despite little literature being available on the use of this method in building physics applications, the authors believe its use is as promising as the PGD approach as it was demonstrated in recent studies [71, 72]. PGD has already been found to be an efficient method, especially because of the great reduction in both computational and storage costs. The comparison of PGD with Spectral methods has been undertaken in [70]. Chapter 9 provided some exercises to introduce the reader to solving diffusion phenomena in building physics by using different numerical methods.

Owing to the evolution of the application of building physics theories for building performance simulation tools that was motivated by energy crises, we believe that, with the progressing concerns associated with climate change and natural disasters, research and development must be focused on the use of advanced numerical methods and investigation of 3D heat and moisture transfer combined with computational fluid dynamics. Furthermore, owing to the dramatic progress in computer hardware, research on the integration of heat and moisture models on a district scale has been emerging, thus inspiring research in urban physics that addresses the urban heat island effect, floods, and air quality, and improves the accuracy of outdoor boundary conditions for a better and more integrated assessment of building energy performance.

Therefore, besides the concerns regarding energy consumption and demand for diffusion phenomena in building physics, it is not hyperbolic to say that the research on the topic presented in this book has social, environmental, and economic aspects.

Finally, the authors hope that this book inspires PhD students to conduct research on building physics and improve the modeling and solutions of diffusion phenomena that occur on a regular basis in buildings in which people spend most of their life time.

References

1. Abadie, M., Mendes, N.: Comparative analysis of response-factor and finite-volume based methods for predicting heat and moisture transfer through porous building materials. J. Build. Phys. **30**(1), 7–37 (2006). https://doi.org/10.1177/1744259106064599
2. Abahri, K., Belarbi, R., Trabelsi, A.: Contribution to analytical and numerical study of combined heat and moisture transfers in porous building materials. Build. Environ. **46**(7), 1354–1360 (2011). https://doi.org/10.1016/j.buildenv.2010.12.020
3. Alamdari, F., Hammond, G.P.: Improved data correlations for buoyancy-driven convection in rooms. Build. Serv. Eng. Res. Technol. **4**(3), 106–112 (1983). https://doi.org/10.1177/014362448300400304
4. Alifanov, O.M., Mikhailov, V.V.: Solution of the nonlinear inverse thermal conductivity problem by the iteration method. J. Eng. Phys. **35**(6), 1501–1506 (1978). https://doi.org/10.1007/BF01104861
5. Allaire, G., Craig, A.: Numerical Analysis and Optimization: An Introduction to Mathematical Modelling and Numerical Simulation. Oxford University Press, Oxford (2007)
6. Allery, C., Béghein, C., Hamdouni, A.: Applying proper orthogonal decomposition to the computation of particle dispersion in a two-dimensional ventilated cavity. Comm. Nonlin. Sci. Num. Sim. **10**(8), 907–920 (2005). https://doi.org/10.1016/j.cnsns.2004.05.005
7. Ammar, A., Mokdad, B., Chinesta, F., Keunings, R.: A new family of solvers for some classes of multidimensional partial differential equations encountered in kinetic theory modelling of complex fluids. J. Non Newtonian Fluid Mech. **144**(2–3), 98–121 (2007). https://doi.org/10.1016/j.jnnfm.2007.03.009
8. Ascher, U.M., Ruuth, S.J., Spiteri, R.J.: Implicit-Explicit Runge-Kutta methods for time-dependent partial differential equations. Appl. Numer. Math. **25**, 151–167 (1997)
9. Barreira, E., Delgado, J.M.P.Q., Ramos, N.M.M., de Freitas, V.P.: Exterior condensations on façades: numerical simulation of the undercooling phenomenon. J. Build. Perform. Simul. **6**(5), 337–345 (2013). https://doi.org/10.1080/19401493.2011.560685
10. Bauklimatik Dresden, B.: Simulation program for the calculation of coupled heat, moisture, air, pollutant, and salt transport. http://www.bauklimatik-dresden.de/delphin/index.php?aLa=en (2011)
11. Beck, J.V., Arnold, K.J.: Parameter Estimation in Engineering and Science. Wiley, New York (1977)
12. Bednar, T., Hagentoft, C.E.: Analytical solution for moisture buffering effect Validation exercises for simulation tools. In: 7th Nordic Symposium on Building Physics. Reykjavik, Iceland (2005)
13. Benjamin, A.T., Walton, D.: Combinatorially composing Chebyshev polynomials. J. Stat. Plan. Inference **140**(8), 2161–2167 (2010). https://doi.org/10.1016/j.jspi.2010.01.012
14. Berger, J.: Contribution à la modélisation hygrothermique des bâtiments: application des méthodes de réduction de modèle. Ph.D. Thesis, Université de Grenoble (2014)

© Springer Nature Switzerland AG 2019
N. Mendes et al., *Numerical Methods for Diffusion Phenomena in Building Physics*,
https://doi.org/10.1007/978-3-030-31574-0

15. Berger, J., Chhay, M., Guernouti, S., Woloszyn, M.: Proper generalized decomposition for solving coupled heat and moisture transfer. J. Build. Perform. Simul. **8**(5), 295–311 (2015). https://doi.org/10.1080/19401493.2014.932012
16. Berger, J., Guernouti, S., Woloszyn, M., Chinesta, F.: Proper generalised decomposition for heat and moisture multizone modelling. Energy Build. **105**, 334–351 (2015). https://doi.org/10.1016/j.enbuild.2015.07.021
17. Berger, J., Mazuroski, W., Mendes, N., Guernouti, S., Woloszyn, M.: 2D whole-building hygrothermal simulation analysis based on a PGD reduced order model. Energy Build. **112**, 49–61 (2016). https://doi.org/10.1016/j.enbuild.2015.11.023
18. Berger, J., Orlande, H.R.B., Mendes, N.: Proper Generalized Decomposition model reduction in the Bayesian framework for solving inverse heat transfer problems. Inverse Probl. Sci. Eng. **25**(2), 260–278 (2017). https://doi.org/10.1080/17415977.2016.1160395
19. Berkooz, G., Holmes, P., Lumley, J.L.: The Proper Orthogonal Decomposition in the Analysis of Turbulent Flows. Ann. Rev. Fluid Mech. **25**(1), 539–575 (1993). https://doi.org/10.1146/annurev.fl.25.010193.002543
20. Bernard, F.: Control System Design: An Introduction to State-Space Methods. Dover Publications Inc, Mineola (2005)
21. Bernardi, C., Maday, Y.: (1989). NASA Report: https://ntrs.nasa.gov/search.jsp?R=19890017257
22. Boyd, J.P.: Chebyshev and Fourier Spectral Methods, 2nd edn. Dover Publications, New York (2000)
23. Brebbia, C.A.: The boundary element method in engineering practice. Eng. Anal. **1**(1), 3–12 (1984). https://doi.org/10.1016/0264-682X(84)90004-2
24. Brebbia, C.A., Dominguez, J.: Boundary Elements: An Introductory Course, 2nd edn. WIT Press, Ashurst (1992)
25. Bueno, A.D.: Heat and Moisture Transfer in Creamic Roof Tile: Simulation and Experimentation. Master, University Federal of Santa Catarina (1994)
26. Burch, D.M.: An analysis of moisture accumulation in walls subjected to hot and humid climates. ASHRAE Trans. **93**(16), 429–439 (1993)
27. Burch, D.M., Thomas, W.C.: An analysis of moisture accumulation in wood frame wall subjected to winter climate. Technical Report, National Institute of Standards and Technology, NISTIR 4674 (1991)
28. Burden, R.L., Faires, J.D., Burden, A.M.: Numerical Analysis, 10th edn. Brooks Cole, Boston (2015)
29. Butcher, J.C.: A history of Runge-Kutta methods. Appl. Numer. Math. **20**(3), 247–260 (1996). https://doi.org/10.1016/0168-9274(95)00108-5
30. Butcher, J.C.: Numerical Methods for Ordinary Differential Equations, 3rd edn. Wiley, Chichester (2016)
31. Caflisch, R.E.: Monte Carlo and quasi-Monte Carlo methods. Acta Numer. **7**, 1–49 (1998). https://doi.org/10.1017/S0962492900002804
32. Cao, S.J., Meyers, J.: Fast prediction of indoor pollutant dispersion based on reduced-order ventilation models. Building Simulation **8**(4), 415–420 (2015). https://doi.org/10.1007/s12273-015-0240-9
33. Cheng, A.H.D., Cheng, D.T.: Heritage and early history of the boundary element method. Eng. Anal. Bound. Elem. **29**(3), 268–302 (2005). https://doi.org/10.1016/j.enganabound.2004.12.001
34. Chetverushkin, B.N., D'Ascenzo, N., Saveliev, V.I.: Hyperbolic type explicit kinetic schemes of magneto gas dynamic for high performance computing systems. Num. Anal. and Math. Model. **30**(1) (2015)
35. Chetverushkin, B.N., Gulin, A.V.: Explicit schemes and numerical simulation using ultrahigh-performance computer systems. Dokl. Math. **86**(2), 681–683 (2012). https://doi.org/10.1134/S1064562412050213
36. Chetverushkin, B.N., Morozov, D.N., Trapeznikova, M.A., Churbanova, N.G., Shil'nikov, E.V.: An explicit scheme for the solution of the filtration problems. Math. Model. Comput. Simul. **2**(6), 669–677 (2010)

37. Cheung, Y.K., Jin, W.G., Zienkiewicz, O.C.: Direct solution procedure for solution of harmonic problems using complete, non-singular. Trefftz functions. Comm. Appl. Num. Meth. **5**, 159–169 (1989)
38. Chhay, M., Dutykh, D., Gisclon, M., Ruyer-Quil, C.: New asymptotic heat transfer model in thin liquid films. Appl. Math. Model. **48**, 844–859 (2017). https://doi.org/10.1016/j.apm. 2017.02.022
39. Chinesta, F., Ammar, A., Leygue, A., Keunings, R.: An overview of the proper generalized decomposition with applications in computational rheology. J. Non-Newtonian Fluid Mech. **166**(11), 578–592 (2011). https://doi.org/10.1016/j.jnnfm.2010.12.012
40. Chinesta, F., Keunings, R., Leygue, A.: The Proper Generalized Decomposition for Advanced Numerical Simulations: A Primer. Springer International Publishing, New York (2013). https://doi.org/10.1007/978-3-319-02865-1
41. Chinesta, F., Leygue, A., Bordeu, F., Aguado, J.V., Cueto, E., Gonzalez, D., Alfaro, I., Ammar, A., Huerta, A.: PGD-based computational vademecum for efficient design, optimization and control. Arch. Comput. Methods Eng. **20**(1), 31–59 (2013). https://doi.org/10.1007/s11831-013-9080-x
42. Cooley, J.W., Tukey, J.W.: An algorithm for the machine calculation of complex Fourier series. Math. Comput. **19**(90), 297–297 (1965). https://doi.org/10.1090/S0025-5718-1965-0178586-1
43. Courant, R., Friedrichs, K., Lewy, H.: Über die partiellen Differenzengleichungen der mathematischen Physik. Math. Ann. **100**(1), 32–74 (1928)
44. Crawley, D.B., Lawrie, L.K., Winkelmann, F.C., Buhl, W.F., Huang, Y.J., Pedersen, C.O., Strand, R.K., Liesen, R.J., Fisher, D.E., Witte, M.J., Glazer, J.: EnergyPlus: creating a new-generation building energy simulation program. Energy Build. **33**(4), 319–331 (2001). https://doi.org/10.1016/S0378-7788(00)00114-6
45. Cunningham, M.J.: The moisture performance of framed structures – A mathematical model. Build. Environ. **23**(2), 123–135 (1988). https://doi.org/10.1016/0360-1323(88)90026-1
46. Cunningham, M.J.: Modelling of moisture transfer in structures — I. A description of a finite-difference nodal model. Build. Environ. **25**(1), 55–61 (1990). https://doi.org/10.1016/0360-1323(90)90041-O
47. Cunningham, M.J.: Modelling of moisture transfer in structures — II. A comparison of a numerical model, an analytical model and some experimental results. Build. Environ. **25**(2), 85–94 (1990). https://doi.org/10.1016/0360-1323(90)90019-N
48. Cunningham, M.J.: Modelling of moisture transfer in structures — III. A comparison between the numerical model SMAHT and field data. Build. Environ. **29**(2), 191–196 (1994). https://doi.org/10.1016/0360-1323(94)90069-8
49. Curran, D.A.S., Lewis, B.A., Cross, M.: A boundary element method for the solution of the transient diffusion equation in two dimensions. Appl. Math. Model. **10**(2), 107–113 (1986). https://doi.org/10.1016/0307-904X(86)90080-6
50. Dauvergne, J.L., Palomo Del Barrio, E.: A spectral method for low-dimensional description of melting/solidification within shape-stabilized phase-change materials. Numer. Heat Transf. Part B Fundam. **56**(2), 142–166 (2009). https://doi.org/10.1080/10407790903116345
51. Dauvergne, J.L., Palomo del Barrio, E.: Toward a simulation-free pod approach for low-dimensional description of phase-change problems. Int. J. Therm. Sci. **49**(8), 1369–1382 (2010). https://doi.org/10.1016/j.ijthermalsci.2010.02.006
52. Delves, L.M., Mohamed, J.L.: Computational Methods for Integral Equations. Cambridge University Press, Cambridge (1985). https://doi.org/10.1017/CBO9780511569609
53. Dormand, J.R., Prince, P.J.: A family of embedded Runge-Kutta formulae. J. Comp. Appl. Math. **6**, 19–26 (1980)
54. Dos Santos, G.H., Mendes, N.: Analysis of numerical methods and simulation time step effects on the prediction of building thermal performance. Appl. Therm. Eng. **24**(8–9), 1129–1142 (2004). https://doi.org/10.1016/j.applthermaleng.2003.11.029
55. Dos Santos, G.H., Mendes, N.: Simultaneous heat and moisture transfer in soils combined with building simulation. Energy Build. **38**(4), 303–314 (2006). https://doi.org/10.1016/j. enbuild.2005.06.011

56. Dos Santos, G.H., Mendes, N.: Combined heat, air and moisture (HAM) transfer model for porous building materials. J. Build. Phys. **32**(3), 203–220 (2009). https://doi.org/10.1177/1744259108098340

57. Dumon, A., Allery, C., Ammar, A.: Proper general decomposition (PGD) for the resolution of Navier-Stokes equations. J. Comp. Phys. **230**(4), 1387–1407 (2011). https://doi.org/10.1016/j.jcp.2010.11.010

58. Einstein, A.: Über die von der molekularkinetischen Theorie der Wärme geforderte Bewegung von in ruhenden Flüssigkeiten suspendierten Teilchen. Annalen der Physik **322**(8), 549–560 (1905). https://doi.org/10.1002/andp.19053220806

59. Einstein, A.: Die Ursache der Mäanderbildung der Flußläufe und des sogenannten Baerschen Gesetzes. Die Naturwissenschaften **14**(11), 223–224 (1926)

60. El Diasty, R., Fazio, P., Budaiwi, I.: Dynamic modelling of moisture absorption and desorption in buildings. Build. Environ. **28**(1), 21–32 (1993). https://doi.org/10.1016/0360-1323(93)90003-L

61. Emmel, M.G., Abadie, M.O., Mendes, N.: New external convective heat transfer coefficient correlations for isolated low-rise buildings. Energy Build. **39**(3), 335–342 (2007). https://doi.org/10.1016/j.enbuild.2006.08.001

62. Evans, L.C.: Partial Differential Equations, 2nd edn. American Mathematical Society, Providence (2010). ISBN: 978-0-8218-4974-3

63. Favier, J., Kourta, A., Leplat, G.: Control of Flow Separation on a Wing Profile Using PIV Measurements and POD Analysis. In: Morrison, J.F., Birch, D.M., Lavoie, P. (eds.) IUTAM Symposium on Flow Control and MEMS, pp. 203–207. Springer, Netherlands, Dordrecht (2008). https://doi.org/10.1007/978-1-4020-6858-4_24

64. Fogleman, M., Rempfer, D., Lumley, J.L., Haworth, D.: POD analysis of in-cylinder flows. In: Volume 2: Symposia and General Papers, Parts A and B, pp. 1173–1178. ASME (2002). https://doi.org/10.1115/FEDSM2002-31413

65. Fornberg, B.: A Practical Guide to Pseudospectral Methods. Cambridge University Press, Cambridge (1996)

66. Fourier, J.: Théorie analytique de la chaleur. Didot, Paris (1822)

67. Fraunhofer, I.: Wufi. http://www.hoki.ibp.fhg.de/wufi/wufi_frame_e.html (2005)

68. Gao, Y., Roux, J.J., Teodosiu, C., Zhao, L.H.: Reduced linear state model of hollow blocks walls, validation using hot box measurements. Energy Build. **36**(11), 1107–1115 (2004). https://doi.org/10.1016/j.enbuild.2004.03.008

69. Gao, Y., Roux, J.J., Zhao, L.H., Jiang, Y.: Dynamical building simulation: a low order model for thermal bridges losses. Energy Build. **40**(12), 2236–2243 (2008). https://doi.org/10.1016/j.enbuild.2008.07.003

70. Gasparin, S., Berger, J., Dutykh, D., Mendes, N.: Advanced reduced-order models for moisture diffusion in porous media. Transp. Porous Media 1–30 (2018). https://doi.org/10.1007/s11242-018-1106-2

71. Gasparin, S., Berger, J., Dutykh, D., Mendes, N.: Solving nonlinear diffusive problems in buildings by means of a Spectral reduced-order model. J. Build. Perform. Simul. (2018). https://doi.org/10.1080/19401493.2018.1458905

72. Gasparin, S., Dutykh, D., Mendes, N.: A spectral method for solving heat and moisture transfer through consolidated porous media. Submitted pp. 1–30 (2018)

73. Gasparini, R.R.: Analysis of conduction and radiation heat transfer in insulation. Master, Pontifical Catholic University of Parana (2005)

74. Gerald, C.F., Wheatley, P.O.: Applied Numerical Analysis, 7th edn. Pearson, London (2003)

75. Girault, M., Derouineau, S., Salat, J., Petit, D.: Réduction de modèle en convection naturelle par une méthode d'identification. C. R. Mécanique **332**(10), 811–818 (2004). https://doi.org/10.1016/j.crme.2004.06.004

76. Girault, M., Maillet, D., Jean-Raymond, F., Braconnier, R., Bonthoux, F.: Estimation of time-varying gaseous contaminant sources in ventilated enclosures through inversion of a reduced model. Int. J. Vent. **4**(4), 365–379 (2006). https://doi.org/10.1080/14733315.2005.11683715

77. Girault, M., Petit, D.: Resolution of linear inverse forced convection problems using model reduction by the Modal Identification Method: application to turbulent flow in parallel-plate duct. Int. J. Heat Mass Transf. **47**(17–18), 3909–3925 (2004). https://doi.org/10.1016/j.ijheatmasstransfer.2004.04.001

78. Girault, M., Petit, D.: Identification methods in nonlinear heat conduction. Part I: Model reduction. Int. J. Heat Mass Transfer **48**(1), 105–118 (2005). https://doi.org/10.1016/j.ijheatmasstransfer.2004.06.032

79. Girault, M., Petit, D., Videcoq, E.: The use of model reduction and function decomposition for identifying boundary conditions of a linear thermal system. Inverse Probl. Eng. **11**(5), 425–455 (2003). https://doi.org/10.1080/1068276031000118961

80. Gisclon, M.: A propos de l'équation de la chaleur et de l'analyse de Fourier. Le journal de maths des élèves **1**(4), 190–197 (1998)

81. Glaser, H.: Graphisches Verfahren zur Untersuchung von Diffusionsvorglinge. Kalfetechnik **10**(1), 345–349 (1959)

82. Goffart, J., Rabouille, M., Mendes, N.: Uncertainty and sensitivity analysis applied to hygrothermal simulation of a brick building in a hot and humid climate. J. Building Perf. Simul. **10**(1), 37–57 (2017). https://doi.org/10.1080/19401493.2015.1112430

83. Goyal, S., Barooah, P.: A method for model-reduction of non-linear thermal dynamics of multi-zone buildings. Energy Build. **47**, 332–340 (2012). https://doi.org/10.1016/j.enbuild.2011.12.005

84. Griffiths, D.F., Dold, J.W., Silvester, D.J.: Separation of variables. In: Essential Partial Differential Equations, pp. 129–159. Springer International Publishing, Berlin (2015). https://doi.org/10.1007/978-3-319-22569-2_8

85. Grunewald, J.: Diffusiver und konvektiver Stoff- und Energietransport in kapillarporösen Baustoffen. Ph.D, TU Dresden (1997)

86. Grunewald, J., Nicolai, A., Li, H., Zhang, J.: Program for predicting temperature profiles in soils: mathematical models and boundary conditions analyses. In: Proceedings of the 8th International Building Performance Simulation Association, pp. 1–8. IBPSA (2003)

87. Grunewald, J., Nicolai, A., Li, H., Zhang, J.: Modeling of coupled numerical HAM and pollutant simulation — Implementation of VOC storage and transport equations. In: Proceedings for the 12th Symposium for Building Physics. NSBP, Dresden, Germany (2007)

88. Gunes, H.: Low-dimensional modeling of non-isothermal twin-jet flow. Int. Commun. Heat Mass Transf. **29**(1), 77–86 (2002). https://doi.org/10.1016/S0735-1933(01)00326-8

89. Hackbusch, W.: The Concept of Stability in Numerical Mathematics. Springer Series in Computational Mathematics, vol. 45. Springer, Berlin (2014). https://doi.org/10.1007/978-3-642-39386-0

90. Hadamard, J.: Sur les problèmes aux dérivées partielles et leur signification physique, pp. 49–52. Princeton University Bulletin, Princeton (1902)

91. Hagentoft, C.E., Kalagasidis, A.S., Adl-Zarrabi, B., Roels, S., Carmeliet, J., Hens, H., Grunewald, J., Funk, M., Becker, R., Shamir, D., Adan, O., Brocken, H., Kumaran, K., Djebbar, R.: Assessment method of numerical prediction models for combined heat, air and moisture transfer in building components: benchmarks for one-dimensional cases. J. Building Phys. **27**(4), 327–352 (2004). https://doi.org/10.1177/1097196304042436

92. Hairer, E., Lubich, C., Wanner, G.: Geometric Numerical Integration. Spring Series in Computational Mathematics, vol. 31, 2nd edn. Springer, Berlin (2006)

93. Hairer, E., Nørsett, S.P., Wanner, G.: Solving Ordinary Differential Equations: Nonstiff Problems. Springer, Berlin (2009). https://doi.org/10.1007/978-3-540-78862-1

94. Hairer, E., Wanner, G.: Solving Ordinary Differential Equations II. Stiff and Differential-Algebraic Problems. Springer Series in Computational Mathematics, Vol. 14 (1996)

95. Heath, M.T.: Scientific Computing: An Introductory Survey, 2nd edn. Mcgraw-Hill, Boston (1997)

96. Herrera, I.: Boundary methods: development of complete systems of solutions. In: Kawai, T. (ed.) Finite Elements Flow Analysis, pp. 897–906. University of Tokyo Press, Tokyo (1982)

97. Herrera, I.: Boundary Methods: An Algebraic Theory. Pitman (1984)

98. Herrera, I.: Trefftz method: a general theory. Numer. Methods Part. Differ. Equ. **16**(6), 561–580 (2000). https://doi.org/10.1002/1098-2426(200011)16:6<561::aid-num4>3.0.CO;2-V
99. Higham, D.J.: An Algorithmic introduction to numerical simulation of stochastic differential equations. SIAM Rev. **43**(3), 525–546 (2001). https://doi.org/10.1137/S0036144500378302
100. Holmes, P.J., Lumley, J.L., Berkooz, G., Mattingly, J.C., Wittenberg, R.W.: Low-dimensional models of coherent structures in turbulence. Phys. Rep. **287**(4), 337–384 (1997). https://doi.org/10.1016/S0370-1573(97)00017-3
101. Huang, C.H., Yan, J.Y., Chen, H.T.: Function estimation in predicting temperature-dependent thermal conductivity without internal measurements. J. Thermophys. Heat Transf. **9**(4), 667–673 (1995). https://doi.org/10.2514/3.722
102. Janssen, H.: Simulation efficiency and accuracy of different moisture transfer potentials. J. Building Perf. Simul. **7**(5), 379–389 (2014). https://doi.org/10.1080/19401493.2013.852246
103. Jarny, Y., Ozisik, M.N., Bardon, J.P.: A general optimization method using adjoint equation for solving multidimensional inverse heat conduction. Int. J. Heat Mass Transf. **34**(11), 2911–2919 (1991). https://doi.org/10.1016/0017-9310(91)90251-9
104. Jaynes, E.T.: Probability Theory. Cambridge University Press, Cambridge (2003)
105. Jirousek, J., Zieliński, A.P.: Survey of Trefftz-type element formulations. Comput. Struct. **63**(2), 225–242 (1997). https://doi.org/10.1016/S0045-7949(96)00366-5
106. Johnson, C.: Numerical Solution of Partial Differential Equations by the Finite Element Method. Dover Publications, New York (2009)
107. Kaipio, J.P., Fox, C.: The Bayesian framework for inverse problems in heat transfer. Heat Transf. Eng. **32**(9), 718–753 (2011). https://doi.org/10.1080/01457632.2011.525137
108. Kalagasidis, A.S.: HAM-Tools: an integrated simulation tool for heat, air and moisture transfer analyses in building physics. Ph.D, Chalmers University of Technology (2004)
109. Kalagasidis, A.S., Weitzmann, P., Nielsen, T.R., Peuhkuri, R., Hagentoft, C.E., Rode, C.: The international building physics toolbox in simulink. Energy Build. **39**(6), 665–674 (2007). https://doi.org/10.1016/j.enbuild.2006.10.007
110. Karagiozis, A.N.: Advanced hygrothermal modeling of building materials using MOISTURE-EXPERT 1.0. In: Proceedings of the International Particleboard/Composite Materials Symposium, pp. 39–47 (2001)
111. Karhunen, K.: Über lineare Methoden in der Wahrscheinlichkeitsrechnung. Kirjapaino oy. sana, Helsinki (1947)
112. Katsikadelis, J.: Boundary Elements: Theory and Applications, 1st edn. Elsevier, Amsterdam (2002). https://www.elsevier.com/books/boundary-elements-theory-and-applications/katsikadelis/978-0-08-044107-8
113. Kerestecioglu, A., Gu, L.: Theoretical and computational investigation of simultaneous heat and moisture transfer in buildings: evaporation and condensation theory. ASHRAE Trans. **Part I**, 447–454 (1990)
114. Kim, E.J.: Development of numerical models of vertical ground heat exchangers and experimental verification : domain decomposition and state model reduction approach. Ph.D, INSA de Lyon (2011)
115. Kita, E., Kamiya, N.: Trefftz method: an overview. Adv. Eng. Softw. **24**, 3–12 (1995)
116. Kolodziej, J.A., Zielinski, A.P.: Boundary Collocation Techniques and their Application in Engineering, 1st edn. WIT Press, Ashurst (2009)
117. Kreiss, H.O., Scherer, G.: Method of lines for hyperbolic equations. SIAM J. Numer. Anal. **29**, 640–646 (1992)
118. Kuenzel, H.M.: Moisture risk assessment of roof construction by computer simulation in comparison to the standard glaser-method. In: International Building Physics Conference. IBBP, Eindhoven (2000)
119. Kuenzel, H.M., Kiessl, J.: Simultaneous heat and moisture transport in building components: one- and two-dimensional calculation using simple parameters. IRB Verlag, Stuttgart (1995)
120. Ladevèze, P.: Sur une famille d'algorithmes en mécanique des structures. Comptes-rendus des séances de l'Académie des sciences. Série 2, Mécanique-physique, chimie, sciences de l'univers, sciences de la terre **300**(2), 41–44 (1985)

121. Lapeyre, B., Lelong, J.: A framework for adaptive Monte-Carlo procedures. Monte Carlo Methods Appl. **17**(1), 77–98 (2011)
122. Larsson, S., Thomée, V.: Finite difference methods for parabolic problems. In: Partial Differential Equations with Numerical Methods, pp. 129–148. Springer, Berlin (2003). https://doi.org/10.1007/978-3-540-88706-5_9
123. Leblond, C., Allery, C.: A priori space-time separated representation for the reduced order modeling of low Reynolds number flows. Comput. Methods Appl. Mech. Eng. **274**, 264–288 (2014). https://doi.org/10.1016/j.cma.2014.02.010
124. Leblond, C., Allery, C., Inard, C.: An optimal projection method for the reduced-order modeling of incompressible flows. Comput. Methods Appl. Mech. Eng. **200**(33–36), 2507–2527 (2011). https://doi.org/10.1016/j.cma.2011.04.020
125. van Leer, B.: Upwind and high-resolution methods for compressible flow: from donor cell to residual-distribution schemes. Commun. Comput. Phys. **1**, 192–206 (2006)
126. LeVeque, R.J.: Finite Difference Methods for Ordinary and Partial Differential Equations. Society for Industrial and Applied Mathematics, Philadelphia (2007). DOI https://doi.org/10.1137/1.9780898717839
127. Li, Z.C., Lu, T.T., Huang, H.T., Cheng, A.H.D.: Trefftz, collocation, and other boundary methods – A comparison. Numer. Methods Part.L Differ. Equ. **23**(1), 93–144 (2007). https://doi.org/10.1002/num.20159
128. Li, Z.C., Lu, T.T., Huang, H.T., Cheng, A.H.D.: Trefftz and Collocation Methods. WIT Press, Ashurst (2008)
129. Liberge, E., Hamdouni, A.: Reduced order modelling method via proper orthogonal decomposition (POD) for flow around an oscillating cylinder. J. Fluids Struct. **26**(2), 292–311 (2010). https://doi.org/10.1016/j.jfluidstructs.2009.10.006
130. Liesen, R.J.: Development of a Response Factor Approach for Modeling the Energy Effects of Combined Heat and Mass Transfer with Vapor Adsorption in Building Elements. Ph.D., University of Illinois (1994)
131. Litz, L.: Order reduction of linear state-space models via optimal approximation of the nondominant modes. IFAC Proc. **13**(6), 195–202 (1980). https://doi.org/10.1016/S1474-6670(17)64799-2
132. Loève, M.: Probability Theory I. Graduate Texts in Mathematics, vol. 45. Springer, New York (1977). https://doi.org/10.1007/978-1-4684-9464-8
133. Lumley, J.L.: The structure of inhomogeneous turbulent flows. In: Yaglom, A.M., Tatarski, V.I. (eds.) Atmospheric turbulence and radio propagation, pp. 166–178. Nauka, Moscow (1967)
134. Mackerle, J.: A guide to the literature on finite and boundary element techniques and software. Eng. Anal. Bound. Elem. **6**(2), 84–96 (1989). https://doi.org/10.1016/0955-7997(89)90004-0
135. For MATLAB, M.C.T.: v4.3.3.12185. Advanpix LLC., Tokyo, Japan (2017)
136. Melo, L.A.: Advanced modelling of heat, air and moisture (HAM) transfer through porous building elements. Ph.D., Pontifical Catholic University of Parana (2017)
137. Mendes, N.: Models for prediction of heat and moisture transfer through porous building elements. Ph.D. Thesis, Federal University of Santa Catarina - UFSC (1997)
138. Mendes, N., Philippi, P.C.: Multitridiagonal-matrix algorithm for coupled heat transfer in porous media: stability analysis and computational performance. J. Porous Media **7**(3), 193–212 (2004). https://doi.org/10.1615/JPorMedia.v7.i3.40
139. Mendes, N., Philippi, P.C.: A method for predicting heat and moisture transfer through multi-layered walls based on temperature and moisture content gradients. Int. J. Heat Mass Transf. **48**(1), 37–51 (2005). https://doi.org/10.1016/j.ijheatmasstransfer.2004.08.011
140. Mendes, N., Philippi, P.C., Lamberts, R.: A new mathematical method to solve highly coupled equations of heat and mass transfer in porous media. Int. J. Heat Mass Transf. **45**(3), 509–518 (2002). https://doi.org/10.1016/S0017-9310(01)00172-7
141. Mendes, N., Ridley, I., Lamberts, R., Philippi, P.C., Budag, K.: Umidus: a PC program for the prediction of heat and mass transfer in porous building elements. In: IBPSA 99, pp. 277–283. International Conference on Building Performance Simulation, Japan (1999)

142. Mendes, N., Winkelmann, F.C., Lamberts, R., Philippi, P.C.: Moisture effects on conduction loads. Energy Build. **35**(7), 631–644 (2003). https://doi.org/10.1016/S0378-7788(02)00171-8

143. Moore, B.: Principal component analysis in linear systems: controllability, observability, and model reduction. IEEE Trans. Autom. Control. **26**(1), 17–32 (1981). https://doi.org/10.1109/TAC.1981.1102568

144. Myshetskaya, E.E., Tishkin, V.F.: Estimates of the hyperbolization effect on the heat equation. Comp. Math. Math. Phys. **55**(8), 1270–1275 (2015). https://doi.org/10.1134/S0965542515080138

145. Neveu, A., El-Khoury, K., Flament, B.: Simulation de la conduction non linéaire en régime variable: décomposition sur les modes de branche. Int. J. Therm. Sci. **38**(4), 289–304 (1999). https://doi.org/10.1016/S1290-0729(99)80095-7

146. Noor, A.K.: Books and monographs on finite element technology. Finite Elem. Anal. Des. **1**(1), 101–111 (1985). https://doi.org/10.1016/0168-874X(85)90011-3

147. Oliveira, A.A.M., Freitas, D.S., Prata, A.T.: Influence of medium properties in diffusivity of Phillip and de Vries model in unsaturated soils. Technical Report, University of Santa Catarina, Florianópolis, Brasil (1993)

148. Orlande, H.R.B., Fudym, O., Maillet, D., Cotta, R.M.: Thermal Measurements and Inverse Techniques. CRC Press, Boca Raton (2011)

149. Osgood, W.F.: Beweis der Existenz einer Lösung der Differentialgleichung ohne Hinzunahme der Cauchy-Lipschitz'schen Bedingung. Monatshefte für Mathematik und Physik **9**(1), 331–345 (1898). https://doi.org/10.1007/BF01707876

150. Ozisik, M.N.: Heat Conduction, 2nd edn. Wiley-Interscience, New York (1993). http://eu.wiley.com/WileyCDA/WileyTitle/productCd-0471532568.html

151. Ozisik, M.N., Orlande, H.R.B.: Inverse Heat Transfer: Fundamentals and Applications. CRC Press, New York (2000)

152. Palomo del Barrio, E., Raji, S., Duquesne, M., Sempey, A.: Reduced models for coupled heat and moisture transfer simulation in wood walls. JP J. Heat Mass Transf. **10**(1), 1–32 (2014)

153. Park, H.M., Chung, O.Y., Lee, J.H.: On the solution of inverse heat transfer problem using the Karhunen-Loève Galerkin method. Int. J. Heat Mass Transf. **42**(1), 127–142 (1999). https://doi.org/10.1016/S0017-9310(98)00136-7

154. Park, H.M., Jung, W.S.: The Karhunen-Loève Galerkin method for the inverse natural convection problems. Int. J. Heat Mass Transf. **44**(1), 155–167 (2001). https://doi.org/10.1016/S0017-9310(00)00092-2

155. Patankar, S.V.: Numerical Heat Transfer and Fluid Flow. CRC Press, Boca Raton (1980)

156. Pearson, K.: The problem of the random walk. Nature **72**, 294 (1905)

157. Pedersen, C.R.: A Transient Model for Analyzing the Hygrothermal Behavior of Building Constructions. In: Proceedings of the 3rd IBPSA (1991)

158. Perrin, B.: Etude des transferts couplés de chaleur et de masse dans des matériaux poreux consolidés non saturés utilisés en génie civil. Ph.D. Thesis, Université Paul Sabatier de Toulouse (1985)

159. Petit, D., Hachette, R., Veyret, D.: A modal identification method to reduce a high-order model: application to heat conduction modelling. Int. J. Model. Sim. **17**(4), 242–250 (1997). https://doi.org/10.1080/02286203.1997.11760336

160. Peyret (2002). https://www.springer.com/gp/book/9780387952215

161. Philibert, J.: One and a half century of diffusion: Fick, Einstein, before and beyond. In: Kärger, J., Grinberg, F., Heitjans, P. (eds.) Diffusion Fundamentals, pp. 8–17. Leipzig Universtätsverlag, Leipzig (2005)

162. Philip, J.R., De Vries, D.A.: Moisture movement in porous materials under temperature gradients. Trans. Am. Geophys. Union **38**(2), 222–232 (1957). https://doi.org/10.1029/TR038i002p00222

163. PUCPR: Domus. http://www.domus.pucpr.br/ (Pontifical Catholic University of Parana) (2018)

164. Qin, Q.H.: The Trefftz Finite and Boundary Element Method, 1st edn. WIT Press, Ashurst (2000)
165. Raudensky, M., Horsky, J., Krejsa, J., Slama, L.: Usage of artificial intelligence methods in inverse problems for estimation of material parameters. Int. J. Numer. Methods Heat Fluid Flow **6**(8), 19–29 (1996). https://doi.org/10.1108/eb017555
166. Reddy, S.C., Trefethen, L.N.: Stability of the method of lines. Numerische Mathematik **62**(1), 235–267. http://www.springerlink.com/index/J6Q3HR77V3627870.pdf (1992). https://doi.org/10.1007/BF01396228
167. Rode, C., Grau, K.: Whole building hygrothermal simulation model. ASHRAE Trans. **109**(1), 572–582 (2003)
168. Ruge, P.: The complete Trefftz method. Acta Mechanica **78**(3–4), 235–242 (1989). https://doi.org/10.1007/BF01179219
169. Saatdjian, E.: Transport Phenomena: Equations and Numerical Solutions. Wiley, New York (2000)
170. dos Santos, G.H., Mendes, N.: Heat, air and moisture transfer through hollow porous blocks. Int. J. Heat Mass Transf. **52**(9–10), 2390–2398 (2009). https://doi.org/10.1016/j.ijheatmasstransfer.2008.11.003
171. dos Santos, G.H., Mendes, N.: Numerical analysis of passive cooling using a porous sandy roof. Applied Thermal Engineering **51**(1–2), 25–31 (2013). https://doi.org/10.1016/j.applthermaleng.2012.08.046
172. dos Santos, G.H., Mendes, N.: Hygrothermal bridge effects on the performance of buildings. Int. Comm. Heat Mass. Transf. **53**, 133–138 (2014). https://doi.org/10.1016/j.icheatmasstransfer.2014.02.018
173. dos Santos, G.H., Mendes, N.: Numerical analysis of hygrothermal performance of reflective insulated roof coatings. Appl. Therm. Eng. **81**, 66–73 (2015). https://doi.org/10.1016/j.applthermaleng.2015.02.017
174. dos Santos, G.H., Mendes, N., Philippi, P.C.: A building corner model for hygrothermal performance and mould growth risk analyses. Int. J. Heat Mass Transf. **52**(21–22), 4862–4872 (2009). https://doi.org/10.1016/j.ijheatmasstransfer.2009.05.026
175. Saulyev, V.K.: Integration of Parabolic Equations by the Grid Method. Fizmatgiz, Moscow (1960)
176. Scheuing, J.E., Tortorelli, D.A.: Inverse heat conduction problem solutions via second-order design sensitivities and newton's method. Inverse Probl. Eng. **2**(3), 227–262 (1996). https://doi.org/10.1080/174159796088027604
177. Schiesser, W.E.: Method of lines solution of the Korteweg-de vries equation. Comput. Math. Appl. **28**(10–12), 147–154 (1994). https://doi.org/10.1016/0898-1221(94)00190-1
178. van Schijndel, A.W.M.: Integrated modeling of dynamic heat, air and moisture processes in buildings and systems using SimuLink and COMSOL. Build. Simul. **2**(2), 143–155 (2009). https://doi.org/10.1007/s12273-009-9411-x
179. Sempey, A., Inard, C., Ghiaus, C., Allery, C.: Fast simulation of temperature distribution in air conditioned rooms by using proper orthogonal decomposition. Build. Environ. **44**(2), 280–289 (2009). https://doi.org/10.1016/j.buildenv.2008.03.004
180. Shampine, L.F.: ODE solvers and the method of lines. Numer. Methods Part. Differ. Equ. **10**(6), 739–755 (1994). https://doi.org/10.1002/num.1690100608
181. Shampine, L.F., Reichelt, M.W.: The MATLAB ODE Suite. SIAM J. Sci. Comput. **18**, 1–22 (1997)
182. Silveira-Neto, A.: Solucoes Exatas Para Problemas de Transporte Simultaneo de Calor e Massa em Elementos Porosos Unidimensionais Master Dissertation. Ph.D., Federal University of Santa Catarina (1985)
183. Solin, P.: Partial Differential Equations and the Finite Element Method. Wiley, Hoboken (2005)
184. Steeman, H.J., Van Belleghem, M., Janssens, A., De Paepe, M.: Coupled simulation of heat and moisture transport in air and porous materials for the assessment of moisture related damage. Build. Environ. **44**(10), 2176–2184 (2009). https://doi.org/10.1016/j.buildenv.2009.03.016

185. Strang, G.: Accurate partial difference methods II. Non-linear problems. Numerische Mathematik **6**(1), 37–46 (1964). https://doi.org/10.1007/BF01386051

186. Strauss, W.A.: Partial Differential Equations: An Introduction, 2nd edn. Wiley, Hoboken (2007)

187. Tariku, F., Kumaran, K., Fazio, P.: Transient model for coupled heat, air and moisture transfer through multilayered porous media. Int. J. Heat Mass Transf. **53**(15–16), 3035–3044 (2010). https://doi.org/10.1016/j.ijheatmasstransfer.2010.03.024

188. Thomée, V.: From finite differences to finite elements. J. Comp. Appl. Math. **128**(1–2), 1–54 (2001). https://doi.org/10.1016/S0377-0427(00)00507-0

189. Trefethen, L.N.: Spectral Methods in MatLab. Society for Industrial and Applied Mathematics, Philadelphia. http://web.comlab.ox.ac.uk/oucl/work/nick.trefethen/spectral.html (2000)

190. Trefftz, E.: Gegenstück zum ritzschen Verfahren. In: Proceedings of the 2nd International Congress for Applied Mechanics, pp. 131–137. Zürich (1926)

191. Uecker, H.: A short ad hoc introduction to spectral methods for parabolic PDE and the Navier-Stokes equations. Technical Report, Carl von Ossietzky Universität Oldenburg, Oldenburg, Germany (2009)

192. Van Belleghem, M.: Modelling coupled heat and moisture transfer between air and porous materials for building applications. Ph.D., Ghent University (2013)

193. Verner, J.H.: Explicit Runge-Kutta methods with estimates of the local truncation error. SIAM J. Num. Anal. **15**(4), 772–790 (1978)

194. Vértesi, P.: Optimal Lebesgue constant for Lagrange interpolation. SIAM J. Numer. Anal. **27**, 1322–1331 (1990)

195. Videcoq, E., Neveu, A., Quemener, O., Girault, M., Petit, D.: Comparison of two nonlinear model reduction techniques: the modal identification method and the branch eigenmodes reduction method. Numer. Heat Transf. Part B Fundam. **49**(6), 537–558 (2006). https://doi.org/10.1080/10407790500344035

196. Videcoq, E., Petit, D.: Model reduction for the resolution of multidimensional inverse heat conduction problems. Int. J. Heat Mass Transf. **44**(10), 1899–1911 (2001). https://doi.org/10.1016/S0017-9310(00)00239-8

197. Videcoq, E., Quemener, O., Lazard, M., Neveu, A.: Heat source identification and on-line temperature control by a branch eigenmodes reduced model. Int. J. Heat Mass Transf. **51**(19–20), 4743–4752 (2008). https://doi.org/10.1016/j.ijheatmasstransfer.2008.02.029

198. Wang, J., Hagentoft, C.E.: A numerical method for calculating combined heat, air and moisture transport in building envelope components. Nord. J. Build. Phys. **2**, 1–17 (2001)

199. Wintner, A.: On the convergence of successive approximations. American Journal of Mathematics **68**(1), 13 (1946). https://doi.org/10.2307/2371736

200. Woloszyn, M., Rode, C.: Tools for performance simulation of heat, air and moisture conditions of whole buildings. Build. Simul. **1**(1), 5–24 (2008). https://doi.org/10.1007/s12273-008-8106-z

201. Wu, C.G., Liang, Y.C., Lin, W.Z., Lee, H.P., Lim, S.P.: A note on equivalence of proper orthogonal decomposition methods. J. Sound Vib. **265**(5), 1103–1110 (2003). https://doi.org/10.1016/S0022-460X(03)00032-4

202. Xiao, D., Fang, F., Buchan, A.G., Pain, C.C., Navon, I.M., Muggeridge, A.: Non-intrusive reduced order modelling of the Navier-Stokes equations. Comput. Methods Appl. Mech. Eng. **293**, 522–541 (2015). https://doi.org/10.1016/j.cma.2015.05.015

203. Yahia, A.A., Palomo Del Barrio, E.: Thermal systems modelling via singular value decomposition: direct and modular approach. Appl. Math. Model. **23**(6), 447–468 (1999). https://doi.org/10.1016/S0307-904X(98)10091-4

204. Yang, C.Y.: Estimation of the temperature-dependent thermal conductivity in inverse heat conduction problems. Appl. Math. Model. **23**(6), 469–478 (1999). https://doi.org/10.1016/S0307-904X(98)10093-8

205. Yik, F.W.H., Underwood, C.P., Chow, W.K.: Simultaneous modelling of heat and moisture transfer and air-conditioning systems in buildings. In: Proceedings IBPSA Building Simulation 95. IBPSA, Madison, WI, USA (1995)

Index

A

Accuracy, 47, 49, 57
Adams–Bashforth methods, 56, 57, 205
Adams–Moulton methods, 57, 205
Advanpix toolbox, 188
Aliasing error, 179
Alternating Direction Implicit method, 83
Anti-aliasing, 180
A posteriori method, 122
Approximating polynomial, 173
A priori method, 122
A-stability, 51, 62
Avogadro number, 5

B

Backward Euler, 49, 205
Backward Time Centered Space (BTCS), 213, 214
Balanced truncation, 123
Basis, 169
BESTEST, 222
Boundary conditions, 106, 160, 195, 196, 203
Boundary Elements Method (BEM), 153
Boundary Integral Equations Methods (BIEM), 153, 154, 159, 164, 194
Boundary integral formulation, 164
Boundary layer, 26
Boundary Value Problem (BVP), 180
Branch Eigenmodes Reduction (BER), 123, 124
Brownian motion, 2, 5, 103, 198, 199
Building physics, 213

C

Central Limit Theorem, 198
CFL condition, 70, 105, 107, 114, 119, 120
Chebfun, 174
Collocation methods, 161, 176, 181, 186, 187, 197
Collocation points, 187
Consistency, 51, 58, 112
Continuous dependence, 207
Convergence, 47, 51, 60
Crank–Nicolson scheme, 71, 81, 109, 215

D

Delaunay triangulation, 100
DELPHIN, 11, 12, 37
Diffusion coefficient, 2, 4, 168
Diffusion problems, 213
Diffusion term, 219
Direct Trefftz methods, 194
Dirichlet boundary, 72, 93, 95, 98, 104, 163–165, 196, 221
Dirichlet-type boundary condition, 154
Discontinuity, 21, 22
Discrete dynamical system, 107
Discrete Fourier Transform (DFT), 179
Dispersion relation, 117
DOMUS, 11–13, 20, 40, 185
Douglas–Gunn method, 83
Dufort–Frankel method, 111, 119
Dufort–Frankel scheme, 71

E

Energy conservation, 9, 16, 20, 22, 24, 28, 31, 34
Energy simulation, 222

© Springer Nature Switzerland AG 2019
N. Mendes et al., *Numerical Methods for Diffusion Phenomena in Building Physics*,
https://doi.org/10.1007/978-3-030-31574-0

Error estimate, 118
Existence of solutions, 207, 208
Expanded polystyrene, 224
Explicit scheme, 106

F
Fast Cosine Fourier Transform (FCFT), 174
Fast Fourier Transform (FFT), 169, 179
FEM basis functions, 92
Feynman–Kac method, 199, 201
Finite difference, 110
Finite Element Method (FEM), 89, 99, 157
Flux, 155
Forward Euler, 48, 49, 204
Fourier coefficients, 170, 179, 180, 183, 186
Fourier modes, 188
Fourier number, 69
Fourier series, 177–179
Fourier space, 184
Fourier transform, 183

G
Galerkin methods, 162, 176, 181, 182, 197
Gear scheme, 71
Generating function, 192
Ghost node, 214
Green function, 4, 104, 154, 156, 163, 166,
        195, 197

H
HAM-Tools, 12, 39
Hat function, 92, 93
Heat and mass transfer, 218, 225
Heat conduction, 182
Heat content, 218
Heat equation, 105, 107, 110, 111, 168, 182,
        183, 199, 202, 203, 219
Heat flux, 106, 185
Helmholtz equation, 89, 153
Hermite polynomials, 175
Heun method, 55, 206
Historical, 9
Homogeneous boundary conditions, 111
Hyperbolic heat equation, 113, 116
Hyperbolization method, 116

I
Implicit method, 54, 226
Implicit scheme, 107, 111, 218
Indirect Trefftz method, 163, 194, 196

Initial condition, 116
Initial Value Problem (IVP), 199, 204, 206,
        207
Interpolating polynomial, 172, 173
Interpolation points, 171

J
Jacobian, 97
Jacobian matrix, 97

K
Kirchoff's transformation, 75

L
Laplace equation, 160, 164, 196
Laplace operator, 103, 156, 160, 170, 195
Large Original Model (LOM), 122
Latent heat, 21
Leap-frog method, 58
Leap-frog scheme, 108
Least square methods, 176, 197
Lebesgue constant, 172–175
Lebesgue measure, 196
Lebesgue points, 174
Legendre polynomials, 171, 174
Linearization of BC, 30, 34
Linear multi-step methods, 54, 56, 57
Lipschitz condition, 209

M
Marshall truncation, 125
Mass conservation, 15, 17, 24, 28, 31
Mass content, 218
MATCH, 10, 12, 36
Matlab, 215
Method of lines, 204
Midpoint method, 57
MIM method, 133
Modal Identification Method (MIM), 123,
        130
Model order, 122
Model reduction, 121
Modified Trefftz method, 163
Moisture content, 10–13, 15, 16, 20–25, 29–
        33, 36, 37, 39, 103, 213, 224, 225,
        227
Moisture volumetric content, 185
Monte–Carlo methods, 198, 201
Multi-Tridiagonal-Matrix Algorithm (MT-
        DMA), 32–34, 40, 80, 221, 225

N

Neumann boundary, 72, 73, 104, 154, 157, 160, 161, 163–165, 196, 214, 225
Neumann nodes, 158

O

Orthogonal polynomials, 171

P

Parabolic equations, 66
Parabolic PDE, 1
Peaceman–Rachford method, 83
Petrov–Galerkin method, 176
Poisson equation, 89, 153, 154, 160
Predictor-corrector method, 56
Prolongation of solutions, 208
Proper Generalised Decomposition (PGD), 124, 146
Proper Orthogonal Decomposition (POD), 123, 139
Pseudo-spectral methods, 170, 178, 182, 185, 192

R

Ralston method, 206
Recursion relation, 190
Reduced Order Model (ROM), 122
Relative humidity, 224, 227
Robin boundary, 73, 115, 196, 215, 219
Runge function, 169, 171, 172
Runge–Kutta, 45
Runge–Kutta methods, 54, 205–207
Runge phenomenon, 172, 175

S

Saulyev method, 113–115, 119
Schrödinger equation, 198
Smooth function, 168
Specific moisture heat, 185
Spectral coefficients, 180, 181, 193
Spectral element, 174
Spectral methods, 167, 170, 221
Stability, 47, 50, 58

Stability condition, 29, 50, 51, 59, 60, 64, 69–71, 79, 118
Strong formulation, 89

T

Tau–Lanczos methods, 176, 181
Taylor expansion, 52, 54, 57, 60, 68, 85, 86
Tchebyshev expansion, 180, 192
Tchebyshev inequality, 198
Tchebyshev nodes, 171, 172, 175, 197
Tchebyshev points, 181
Tchebyshev polynomials, 171, 174, 190–192, 194
Tchebyshev series, 177
T-complete functions, 159, 160, 164
T-expansion, 160
T-functions, 163
θ-scheme, 71
Thomas algorithm, 73, 82, 83
Trapezoidal rule, 57
Trefftz direct method, 164
Trefftz expansion, 163
Trefftz indirect method, 160
Trefftz method, 153, 194
Tridiagonal matrix, 215
Tridiagonal matrix algorithm, 73
Trigonometric basis, 170
Truncated expansion, 192

U

Uniqueness of solutions, 207–209

V

Vapor flow, 21
von Neumann stability, 70, 112

W

Weak formulation, 90
WUFI, 11, 12, 22, 32, 38, 39

Z

Zero-stability, 51, 58

Printed in the United States
By Bookmasters